T0258110

Biomaterials: Diverse Applications

Biomaterials:
Diverse Applications

Edited by **Ralph Seguin**

New York

Published by NY Research Press,
23 West, 55th Street, Suite 816,
New York, NY 10019, USA
www.nyresearchpress.com

Biomaterials: Diverse Applications
Edited by Ralph Seguin

© 2015 NY Research Press

International Standard Book Number: 978-1-63238-063-0 (Hardback)

Printed in the United States of America.

Contents

            Permissions

            List of Contributors

Preface

This book is a collection of reviews and original researches by experts and scientists regarding the field of biomaterials, its development and applications. It offers readers the potentials of distinct synthetic and engineered biomaterials. This book gives a comprehensive summary of the applications of various biomaterials, along with the techniques required for designing, developing and characterizing these biomaterials without any intervention by any industrial source. The researches and reviews focus on the functions of novel and familiar macromolecular compounds in nanotechnology and nanomedicine. The book also elucidates the chemical and mechanical modeling of these compounds so as to serve biomedical needs.

Various studies have approached the subject by analyzing it with a single perspective, but the present book provides diverse methodologies and techniques to address this field. This book contains theories and applications needed for understanding the subject from different perspectives. The aim is to keep the readers informed about the progresses in the field; therefore, the contributions were carefully examined to compile novel researches by specialists from across the globe.

Indeed, the job of the editor is the most crucial and challenging in compiling all chapters into a single book. In the end, I would extend my sincere thanks to the chapter authors for their profound work. I am also thankful for the support provided by my family and colleagues during the compilation of this book.

Editor

New and Classical Materials
for Biomedical Use

1

Nacre, a Natural Biomaterial

Marthe Rousseau
Henri Poincaré University, Nancy I
France

1. Introduction

Nacre, or mother-of-pearl, is a calcium carbonate structure produced by bivalves, gastropods, and cephalopods as an internal shell coating. Because of its highly organized internal structure, chemical complexity, mechanical properties and optical effects, which create a characteristic and beautiful lustre, the formation of nacre is among the best-studied examples of calcium carbonate biomineralization. In this chapter we will detail the structure of nacre and its growth mechanism. We will summarize the several results obtained on the *in vivo* and *in vitro* biological activity of nacre and its organic molecules.

2. Nacre structure

The interdigitating brickwork array of nacre tablets (Fig.1), specific, in bivalves ("sheet nacre") is not the only interesting aspect of nacre structure. Nacre is an organo-mineral composite at micro- and nano-scale. The bio-crystal itself is a composite. It has not only the mineral structure of aragonite but possesses intracrystalline organic material (Watabe, 1965). The primary structural component is a pseudohexagonal tablet, about 0.5 µm thick and about 5 to 10 µm in width, consisting primarily (97%) of aragonite, a polymorph of $CaCO_3$, and of organics (3%). Nacre can be worked as pieces and powder.

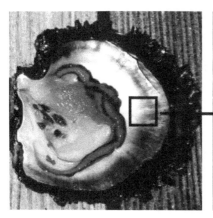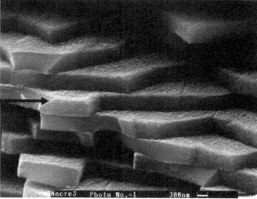

Fig. 1. Characteristic brick and mortar structure of the nacreous layer of *Pinctada*.

Fig. 2. Darkfield TEM image of nacre evidencing the crystalline structure of the organic matrix (bar is 100nm). Organic matter is in contrast when under Bragg conditions whilst the mineral phase remains systematically extinguished.

Transmission electron microscopy performed in the darkfield mode evidences (Fig.2) that intracrystalline matrix is highly crystallised and responds like a 'single crystal'. The organic matrix is continuous inside the tablet, mineral phase is thus finely divided but behaves in the same time as a single crystal.

The tablet of nacre, the biocrystal, does diffract as a single crystal but is made up of a continuous organic matrix (intracrystalline organic matrix) which breaks the mineral up into coherent nanograins (~45 nm mean size, flat-on) which share the same crystallographic orientation. The single crystal-like mineral orientation of the tablet is supposedly created by the heteroepitaxy of the intracrystalline organic matrix. This is a strong hypothesis because this work demonstrates at the same time that this intracrystalline matrix is well crystallized (i.e. periodic) and diffracts as a "single crystal" too (darkfield TEM mode, Fig. 2). However these "organic crystals" do not show the same orientation in adjacent tablets.

Neighbouring tablets above and below can maintain a common orientation, which again raises the issue of the transmission of mineral orientation from one row to the next. Bridges are well identified in *Pinctada* between successive rows in the pile. This implies that an organic template is controlling the orientation of the aragonite.

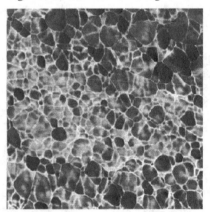

Fig. 3. AFM picture in Phase Contrast (1x1μm²) at the nanometer (nm) length scale, the aragonite component inside individual tablets is embedded in a crystallographically oriented foam-like structure of intra-crystalline organic materials in which the mean size of individual aragonite domains is around 50 nm.

Intermittent-Contact Atomic Force Microscopy with phase detection imaging reveals a nanostructure within the tablet (Fig.3). A continuous organic framework divides each tablet into nanograins. Their mean extension is 45nm. It is proposed that each tablet results from the coherent aggregation of nanograins keeping strictly the same crystallographic orientation thanks to a hetero-epitaxy mechanism (Rousseau et al., 2005a).

3. Nacre growth mechanism

Formation of nacre (*mother-of-pearl*) is a biomineralization process of fundamental scientific as well as industrial importance. However, the dynamics of the formation process is still not all understood. Scanning electron microscopy and high spatial resolution ion-microprobe depth-profiling have been used to image the full three-dimensional distribution of organic materials around individual tablets in the top-most layer of forming nacre in bivalves. Nacre formation proceeds by lateral, symmetric growth of individual tablets mediated by a growth-ring rich in organics, in which aragonite crystallizes from amorphous precursors (Fig.4). The pivotal role in nacre formation played by the growth-ring structure documented in this study adds further complexity to a highly dynamical biomineralization process.

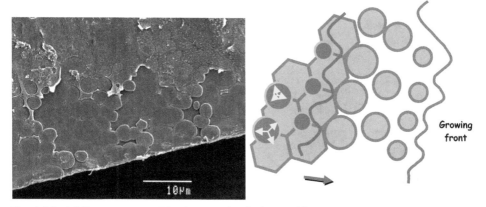

Fig. 4. Nuclei (future nacre tablets) grow circularly in a film, get in contact and form polygons.

Figure 5 shows the surface membrane, observed from the point of view of the nacre layer, and corresponding features in the growing nacre surface, respectively. Several important observations can be made. i) For each nacre tablet, there exists in the membrane a circular structure (Fig. 5a, c). The circular membrane structure displays higher topography than the adjacent surface membrane. We refer to this circular structure as the 'ring'. ii) This ring makes direct physical contact with the growing nacre tablets. Each nacre tablet features a similar ring, several hundred nm wide, with a complementary (i.e. lower) topography (Fig. 5b, d). iii) Neighbouring tablets can merge during growth by lateral expansion that leads to a 'collision'. In this process, ring structures surrounding each tablet first merge together, then disappear (Fig. 5d).

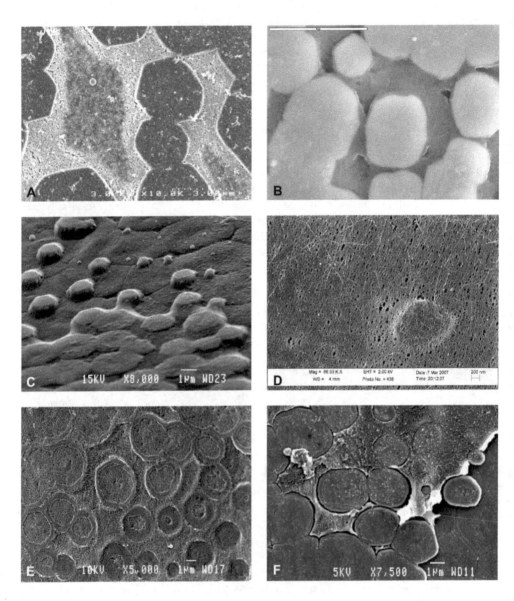

Fig. 5. Direct observation of the nacre surface of the same sample series with different SEM secondary electron detectors : A: in-lens detector at 3kV, B: same but at 15kV both with an Hitachi 4500-FEG microscope, C, E, F: lower detector of the JEOL JSM-840A at 15, 10 and 5kV, D: with a high-resolution Ultra 55 Zeiss using an in-lens secondary electron detector at 2kV.

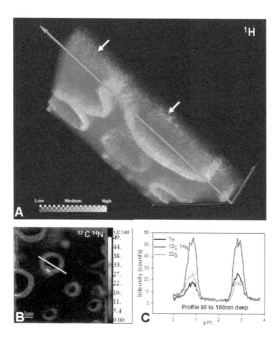

Fig. 6. Distribution of organics around nacre tablets. a) A 3-dimensional reconstruction of the H distribution at different depths in the surface layers of the forming nacre. The data were obtained with the NanoSIMS ion microprobe. The direction of sputtering is indicated by a white arrow. The sputtered volume is about 10μmx10μmx0.2μm. For greater clarity, the thickness of this volume (blue axis) has been expanded by a factor of 6. The surface membrane is visible in the first images (white arrow) but deeper in the structure organics are concentrated only in the growth-rings surrounding each nacre tablet. (Full 3-D reconstructions of the H, N and S distributions are available in SD as movies) b) NanoSIMS image showing the distribution of N, which is also preferentially concentrated in the growth-ring around each nacre tablet. c) Line-profile (track indicated in b) showing the enhanced concentrations of H, N, and S in the growth-ring. See text for discussion.

The NanoSIMS allows chemical composition to be imaged with extremely high spatial resolution (on the order of 100 nm), while depth-profiling through the inter-lamellar matrix and into the nacreous structure with a depth resolution of about 10 nm per image (Meibom et al., 2004, 2007). The 3-dimensional distribution of H, N and S are diagnostic of organic compounds in which their concentration is greatly increased over that in the aragonitic nacre tablets. Figure 6a shows the 3D reconstruction of a depth-sequence of H images from the top-most layers (in the membrane), via intermediate levels to an estimated depth of around 0.2 μm in the tablet structure, where the overlying membrane has been sputtered away completely and the distribution of organics around individual tablets becomes visible. The distribution of organics in the ring surrounding each nacre tablet corresponds to the characteristic ring in the overlying mantle; see Fig. 5d. Figure 6c shows in greater detail the distribution of organics around individual tablets. Importantly, concentrations of H, N and S are clearly confined to the rim around each tablet. In the ring, the signal from N is almost two orders of magnitude higher than in the surrounding aragonitic tablets. Sulfur and Hydrogen are enriched by factors of about 10 and 7, respectively over the signal observed from inside the tablet. We proposed a model illustrated in Fig. 7. A growth-ring structure, rich in organic materials, surrounds each growing nacre tablet during formation of sheet nacre in *Pinctada margaritifera* bivalves. This structure disappears as nacre tablets grow laterally and collide with adjacent tablets. It is conceivable that this organic ring structure acts to nucleate aragonite into the highly oriented nano-crystals (~50 nm in size) that make up the meso-crystal (i.e. µm sized) nacre tablets (Wohlrab et al., 2005; Kulak et al., 2007). This adds support for a highly dynamic biomineralization process during which organic materials and carbonate precursor phases (likely amorphous) are delivered to the site of nacre tablet formation from the overlying mantle with a high degree of spatial control.

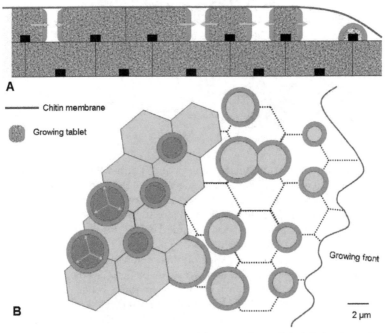

Fig. 7. Schematic representation of the nacre tablet growth-model. a) Side-view of the top-most tablet-layers. Underneath the top membrane, which is in direct contact with the overlying mantle of the animal, individual tablets grow laterally from a carbonate-charged silk-phase through crystallization mediated by organics in the nucleation sites (black rectangles) and in the growth-rings (shown in red). The nucleation sites are generally placed above the junctions of three tablets in the underlying layer of tablets. When two tablets collide during growth, the growth-rings first merge and subsequently disappear, allowing adjacent tablets to make physical contact. b) Top view of the nacre growth front. Individual tablets nucleate and grow concentrically reflecting the shape of the surrounding organic growth-ring. Only after collision and disappearance of the growth-rings do individual tablets take the form of hexagons. The hexagonal geometry of individual tablets is therefore the result of the distribution of nucleation sites, which inherit their distribution from the underlying layer.

4. Biological action of raw nacre

Nacre biological action. A major breakthrough was done in 1992, when E. Lopez et al. discovered that natural nacre from the pearl oyster *Pinctada maxima* is simultaneously biocompatible and osteoinductive. Nacre shows osteogenic activity after implantation in human bone environment (Silve et al., 1992). Raw nacre pieces designed for large bone defects were used as replacement bone devices in the femur of sheep. Over a period of 12 months, the nacre blocks show persistence without alteration of the implant shape. A complete sequence of osteogenesis resulted from direct contact between newly formed bone and the nacre, anchoring the nacre implant (Atlan et al., 1997). Furthermore, when nacre is implanted in bone, new bone formation occurs, without any inflammatory reaction and fibrous formation. We observed an osteoprogenitor cellular layer lining the implant, resulting in a complete sequence of new bone formation (Fig.8). Results showed calcium and phosphate ions lining the nacre within the osteoprogenitor tissue (Atlan et al, 1999).

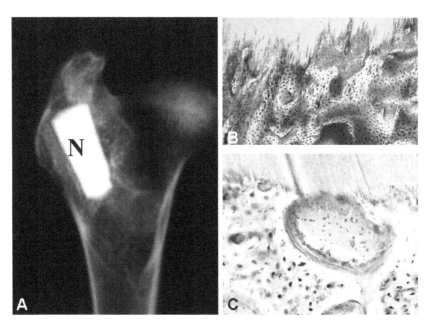

Fig. 8. Radiographic image of the femur epiphysis harvested 10 months after implantation of a block of nacre (N) showing the preserved size and shape of the implant and the close contact between the implant and surrounding cancellous bone (A). After erosion of the implant surface nacre is in direct contact with the new formed bone (B). At the interface bone forming cells are stimulated (C) (Atlan et al, 1999).

Osteogenesis thought to begin with the recruitment of mesenchymal stem cells, which differentiate to form osteoblasts in response to one or more osteogenic factors. In previous studies (Atlan et al., 1997; Silve et al., 1992), as reviewed in Westbroek and Marin (1998), it has been shown by means of *in vivo* and *in vitro* experiments that nacre can attract and activate bone marrow stem cells and osteoblasts.

Other authors have demonstrated the same activity of nacre. Liao et al., in 2000 have published results about the implantation in back muscles and femurs of rats of nacre pieces coming from the shell of the freshwater *Margaritifera*. They concluded that their nacre was biocompatible, biodegradable and osteoconductive material. They confirmed the results obtained with *Pinctada* nacre. More recently Shen et al. (2006) have demonstrated the *in vitro* osteogenic activity of pearl. Hydroxyapatite can be formed on pearl surface in Simulated Body Fluid based on a dissolution-binding-precipitation mechanism. Cell culture shows that pearl has the same osteogenic activity as shell nacre.

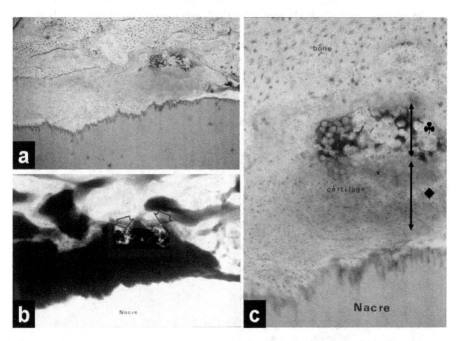

Fig. 9. Histological study of the nacre/ bone interface, lateral to the nacre trochlea, after 6 months of implantation: longitudinal section of nacre subchondral implant, islets of ossification at the interface between nacre and bone (a: x8; b: microradiography of the same area), showing the sequence of cell differentiation from nacre to spongy host bone, chondrocytes (♦) and hypertrophic chondrocytes (♣), after basic fuchsin and toludine bleu staining (c: x 16).

Nacre has also been implanted in the subchondral bone area in the sheep knee. We implanted nacre blocks in sheep trochlea by replacing the half of the femoral trochlea (nacre group). For comparison we used complete cartilage resection (resection group) down to the subchondral bone. In the "nacre group", implants were well tolerated without any synovial inflammation (Fig.9). This complementary study was the first one designed to analyze the behaviour of nacre *in vivo*, in the intra articular cavity after implantation into the subchondral bone in the sheep's knee (Delattre, 2000). After 6 months of implantation, a new tissue was formed on the articular surface of nacre. This tissue is composed of new formed bone and articular cartilage, partially covering the subchondral implant placed in the defect. The presence of nacre can progressively stimulate the regrowth of a tissue, reproducing the functional osteochondral structure in which the subchondral bone sends stimulating messages to the cartilage layer. The regrowth reaction seems to be a well regulated physiological process, demonstrating the efficacy and good tolerance to nacre. Laterally to the nacre implant we observed islets of endochondral ossification. Endochondral ossification is initiated by the formation of cartilage templates of future bones, built by mesenchymal progenitor cells, which condensate and differentiate into chondrocytes. During the repair process the events of endochondral ossification are recapitulated (Fig.9).

A process of nacre powder preparation has been patented (Lopez et al., 1995). The obtained nacre powder has been experimented in injectable form in vertebral and maxillar sites of sheep (Lamghari et al., 1999a, 1999b). This work has demonstrated that nacre powder is resorbable and that this resorption induced the formation of normal bone. The nacre powder filled the whole experimental cavity after 1 week post-surgery. There was no inflammatory or foreign body reaction in the cavity area. Samples taken at 8 weeks after injection showed dissolution of the nacre within the cavity. Angiogenesis had began by that time and the cavity was invaded by a network of capillaries. The cavity contained newly formed woven bone (Fig.10). The vertebral bone adjacent to the cavities contained interconnected bone lamellae. They were also bone remodelling units with central lacunae rich in bone marrow. This new formed Bone was functional as normal bone.

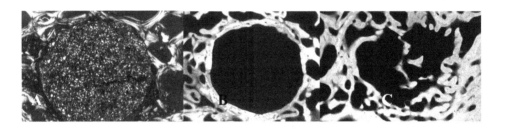

Fig. 10. Nacre injected into the sheep vertebrae. Polarized light image showing transverse section throught a bone defect. The cavity is filled with nacre (A, x25). In the empty cavity after 8 weeks, the bone was organised into concentric rings (B). 8 weeks after injection the nacre powdered has disappeared and been replaced by new formed bone (C) (Lamghari et al, 2001).

In the *in vivo* studies nacre has been shown to be resorbable. Osteoclast activity was therefore studied *in vitro* on nacre (Fig.11). The idea was to assess the plasticity of bone resorbing cells and their capacity to adapt to a biomineralized material with a different organic and mineral composition. Osteoclast stem cells and mature osteoblasts were cultured on nacre substrate and osteoclast precursor were shown to differentiate into osteoclasts capable of resorbing nacre (Duplat et al, 2007).

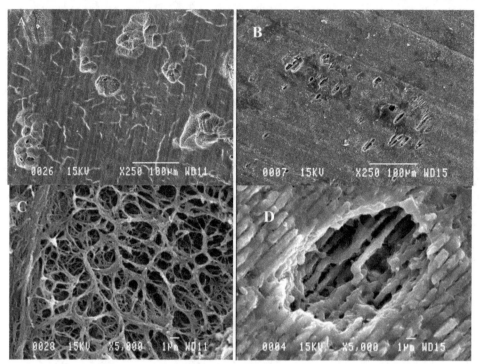

Fig. 11. Resorption process of osteoclasts differentiated from human CD14+ monocytes on bone (A, C) and on nacre (B, D). SEM examination of resorption lacunae. The form of the lacunae is quite different and can be explained by the different organic and mineral compositions as well as by the mineral density of bone and nacre.

5. Biological action of nacre extracts

Nacre is composed of aragonite crystal tablets covered by an organic matrix (3%) (Bevelander & Nakahara, 1969). This organic phase is composed of chitine, polysaccharides, proteins, peptides, lipids and other small molecules lower than 1000 Da. The organic molecules can be extracted with aqueous and organic solvents. The water-soluble organic matrix dictates which calcium carbonate crystal structure is formed and when it is deposited (Cariolou & Morse, 1988). The bulk of the watersoluble biopolymer is thought to consist of a complex mixture of proteins and peptides (Kono et al., 2000; Weiss et al., 2000). Nacre possess in its organic matrix molecular signals that have the ability to trigger bone cell commitment. Efforts were undertaken in order to identify these active molecules and to determine their mode of action on mammalian bone cells. A particular attention was paid to the water-soluble matrix (WSM) molecules of nacre. We supposed that this fraction contains molecules which are released when the nacre implants are placed in living systems. Molecules extracted from nacre of the pearl oyster *P. maxima* with water have been shown to have a biological activity on various mammalian pre-osteogenic cell types (Rousseau et al., 2003).

The activity of the nacre water-soluble matrix (WSM) from *P. maxima* was measured on the clonal osteogenic cell-line MC3T3-E1 established from newborn mouse calvaria. These cells have the capacity to differentiate into osteoblasts, and form calcified bone tissue *in vitro*. The WSM increased alkaline phosphatase activity (Pereira-Mouriès et al., 2002b) and induced the formation of bone nodules (Rousseau et al., 2003). On bone marrow stromal cells, WSM stimulated the proliferation, the differentiation and the early mineralization of osteoprogenitor cells (Lamghari et al., 1999a).

Recently, a great number of low molecular weight molecules was identified in WSM as main components, whereas proteins are minor (15 % w/w). It was supposed that the signal-molecules of nacre might to be low molecular weight molecules. This feature may facilitate their diffusion in the host tissues from nacre implants. The hypothesis that nacre low MW molecules are active in bone cell differentiation was evaluated. Water-soluble molecules from nacre were fractionated according to dialysis, solvent extraction and reversed-phase HPLC. The two sub-fractions ESM and F1 were tested on MC3T3-E1 cell culture. Mass spectrometry analysis showed that the F1 sub-fraction contains around 30 polar molecules ranging from 50 to 300 Da (Fig.12). Peptides were not identified in this fraction. However, the aggregative property of these molecules during chromatography precludes the obtaining of a more purified active fraction.

The presence of BMP-like molecules in WSM was supposed but not demonstrated (Rousseau et al., 2007). Molecules isolated from nacre induced mineralization of the preosteoblasts extracellular matrix after 16 days of culture that was analyzed as hydroxyapatite by Raman spectroscopy. This study indicated that the nacre molecules efficient in bone cell differentiation are certainly different from proteins, and probably more related to peptides (Rousseau et al., 2003, 2007). Molecules isolated from nacre, ranging from 50 to 235 Da, in induced red alizarin staining of the preosteoblasts extracellular matrix after 16 days of culture. The treatment of cells with nacre molecules accelerated expression of collagen I and increased mRNA expression of Runx2 and osteopontin (Fig.13). Raman spectroscopy demonstrated the presence of hydroxyapatite in samples treated with this molecules. Scanning electron microscopy pictures showed at the surface of the treated cells the occurrence of clusters of spherical particles resembling to hydroxyapatite (Fig. 14). Nacre low molecular weight molecules stimulate the early stages of bone differentiation and the formation of an extracellular matrix able to initiate hydroxyapatite nucleation.

The water-soluble matrix of nacre is composed of at least 110 molecules ranging from 100 to 700 Da, with a low content (10 %) of peptides (Bédouet et al., 2006). On the other hand, the protein content of the WSM represents 15 % (w/w) of the extract (Pereira-Mouriès et al., 2002a). Dialysis fractionation of the water-soluble matrix of nacre indicated that low molecular mass molecules lower than 1kDa were important components. They represents 0.14% of the nacre weight, i.e. 60% of the water soluble matrix itself, and contained a huge chemical diversity, since 100 molecules with close molecular weight were gathered after MS analyses of only half of the recovered fractions lower than 1kDa. Some of the low molecular weight molecules were glycine-rich pepetides (9%) whereas the chemical nature of the others remains unidentified. These molecules could correspond either to degradation product of biopolymers or to metabolites accumulated in nacre during the growth of the oyster shell.

The low molecular weight fraction contained specific inhibitors of cysteine protease in addition to proteinase K inhibitors (Bédouet et al., 2007). The specificity of the proteinase inhibitors Found in the nacre water-soluble fraction is quite interesting. Indeed, the cysteine proteinases, particularly cathepsin B and L, are involved in different pathologies, such as arhtrosis and cartilage degradation. The organic matrix of nacre contains both general and more specific inhibitors of cysteins protease in addition to proteinase K inhibitors. The mollusc shell may be used as a new source of natural inhibitors, particularly for the cystein proteinases. The proteinase inhibitors present in the shell may play a major role in the regulation of biomineralization, and they may protect the nacre layer against proteolytic digestion during the perforation of the shell by worms.

A

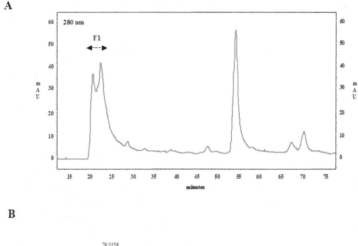

B

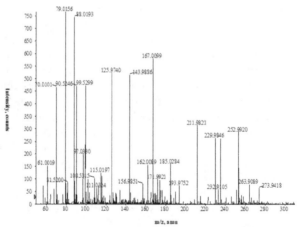

Fig. 12. Isolation of active fraction F1 from the nacre low molecular weight molecules using reversed-phase chromatography. The ESM fraction (75 mg) was loaded onto a C₄ column (2.2 x 25 cm) before elution with a linear acetonitrile-0.1 % TFA gradient. The flow rate was 3 mL/min and fractions (5 mL) eluted between 16 and 25 min were pooled giving the fraction F1 (A). Analysis using electrospray mass spectrometry (ion-positive mode) of the fraction F1. Labels indicate [M+H$^+$]$^+$ ion species (B).

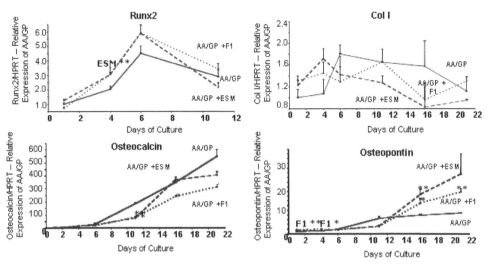

Fig. 13. Effects of ESM and F1 fractions on Runx2, type I collagen, osteopontin and osteocalcin gene expression in MC3T3-E1 cells analyzed by quantitative RT-PCR. The MC3T3-E1 cells were cultured during 1, 4, 6, 11, 16 and 21 days in the absence or presence of ESM and F1 fraction (200 µg/mL) in α-MEM medium supplemented with 10 mM GP (β-glycerophosphate) and 50µg/µL AA (ascorbic acid). Total cellular RNA was isolated and expression of Runx2, type I collagen, osteopontin and osteocalcin were determined by quantitative RT-PCR relative to the HPRT expression. Results are the means + SEM of 5 experiments for Runx2 and 3 experiments for col I, osteopontin and osteocalcin. The reference is day 1 from AA/GP samples. Data were corrected for initial template quantity variations by normalizing values by HPRT. A minimum significant difference (* p<0.05, ** p<0.01) between control and the samples ESM and F1 was determined.

Lipids presence was discovered in the so-called "ethanol soluble matrix" (ESM). The extract contains fatty acids, triglycerides, cholesterol and ceramides in low abundance (Fig.15). Application of ESM on human skin explants previously partially dehydrated induced an overexpression of filaggrin (responsible for hydration of stratum corneum) and a decrease of transglutaminase expression (overexpressed in inflammatory skin diseases). We also showed that ESM extracted from the mother of pearl of *P. margaritifera* induced a reconstitution of the intercellular cement of the *Stratum corneum* (Fig.16) (Rousseau et al., 2006). If the physiological and structural functions of the lipids in the nacre remain elusive, their biological activity on delipidated human skin explants is more accurate. Two lipids preparations containing respectively 0.5 and 1% of nacre lipids in weight were prepared with an excipient. According to the immunolabelling of filaggrin in skin explants, it has been showed that the nacre lipids induced on the delipidated skin an expression of filaggrin higher to the the level found in untreated skin. On the other hand, the lipid formulation reduced the expression of TGase 1 in the cornified envelope in a dose dependent manner. It seemed that the lipids extracted from nacre had brought to the stratum corneum the elements necessary to a rapid reconstruction of the intercellular cement, by acting on the expression levels of transglutaminase and filaggrin.

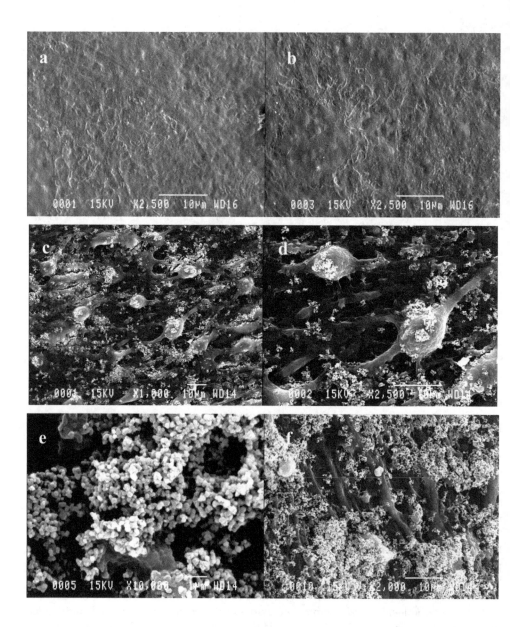

Fig. 14. Scanning electron microscopy observation of the MC3T3-E1 cells extracellular matrix treated with ESM for 16 days. Control cell (a), cells treated with AA/GP (b), cells treated with ESM (200 µg/mL) in presence of AA/GP (c, d, e, f) at different magnifications.

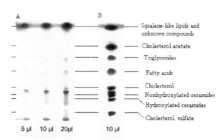

Fig. 15. TLC analysis of lipids extracted from the nacre of *P. margaritifera* with hexane/diethylether/acetic acid (60,25,15, v/v/v) as a mobile phase. Samples of 5, 10 and 20 µl of the lipid extracts at 20 mg/ml and a standard lipid mixture loaded and stained with copper sulfate reagent.

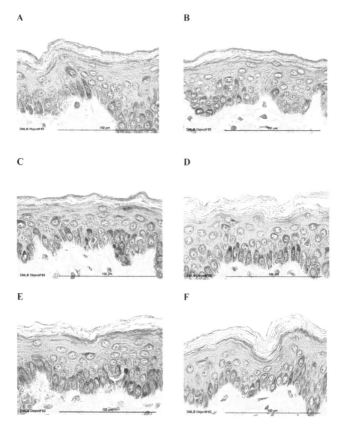

Fig. 16. Histological observation of the skin explants morphology. Control at To (A), in dehydrated skin at To (B), in dehydrated skin at T + 3 h (C), in dehydrated skin at T + 3 h treated with excipient (D) and in dehydrated skin treated with 0.5 % (E) and 1 % (F) of nacre lipids at T + 3 h. Staining was performed according to the Masson's Trichrom.

6. Conclusion

Nacre, or mother-of-pearl, is a calcium carbonate structure produced by bivalves, gastropods, and cephalopods as an internal shell coating. Because of its highly organized internal structure, chemical complexity, mechanical properties and optical effects, which create a characteristic and beautiful lustre, the formation of nacre is among the best-studied examples of calcium carbonate biomineralization. A major breakthrough was done in 1992, when E. Lopez et al. discovered that natural nacre from the pearl oyster *Pinctada* is simultaneously biocompatible and osteoinductive. A complete sequence of osteogenesis resulted from direct contact between newly formed bone and the nacre, anchoring the nacre implant. Furthermore, when nacre is implanted in bone, new bone formation occurs, without any inflammatory reaction and fibrous formation. In previous studies as reviewed in Westbroek and Marin, it has been shown by means of *in vivo* and *in vitro* experiments that nacre can attract and activate osteoblasts. Nacre has been implanted in the subchondral bone area in the sheep knee. A process of nacre powder preparation has been patented. The obtained nacre powder has been experimented in injectable form in vertebral and maxillar sites by sheep. Nacre implant induced no inflammatory reaction. Nacre is biocompatible, osteogenic and osteoinductor.

Nacre is composed of aragonite crystal tablets covered by an organic matrix. The water-soluble organic matrix dictates which calcium carbonate crystal structure is formed and when it is deposited. The bulk of the watersoluble fraction is thought to consist of a complex mixture of proteins and peptides.

The water soluble matrix (WSM) increases alkaline phosphatase activity and induces formation of bone nodules of the clonal osteogenic cell-line MC3T3-E1. WSM, Dexamethasone and BMP-2 all stimulate alkaline phosphatase activity in a manner corresponding to a differentiation into an osteoblast phenotype. On bone marrow stromal cells, WSM stimulates the proliferation, the differentiation and the early mineralization of osteoprogenitor cells. Small molecules isolated from nacre induce mineralization of the preosteoblast extracellular matrix after 16 days of culture. Raman spectroscopy revealed as hydroxyapatite the crystals formed extracellularly. We also extracted proteinase inhibitors from the nacre of *Pinctada*. The low molecular weight fraction contained specific inhibitors of cysteine protease in addition to proteinase K inhibitors. We also discovered the presence of lipids in nacre and we showed that their lipids induced a reconstitution of the intercellular cement of the human *Stratum corneum*.

Nacre has been mostly studied the last two decades for its action on bone and skin (Table 1).

Biocompatiblity
-   Nacre is not toxic for the human body after maxillary implantation
-   When nacre is implanted in bone, no inflammatory reaction nor fibrous formation was observed.

Effects on Bone
-   Osteogenic activity was observed in sheep after implantation in bone environment
-   Nacre powder is resorbable and this resorption induced the formation of normal bone *in vivo*
-   Nacre can attract and stimulate osteoblasts activity *in vitro*
-   The nacre molecules involved in bone cell differentiation are small, certainly different from proteins, and probably peptides

Effects on Cartilage
-   Nacreous trochlea is covered with new non fibrous cartilage after implantation in sheep

Anti-proteases effect
-   Presence of cysteine proteases and proteinase K inhibitors in nacre organics

Effect on Skin
-   Nacre lipids are able to reconstitute the epiderm intercellular cement

Table 1. Summary of the biological actions of nacre

7. Acknowledgement

We would like to thank all the investigators and co-investigators involved in the research on nacre: particularly Professor Evelyne Lopez for the discovery of the biological activity on nacre, Dr Xavier Bourrat for his help on the structure study and the growth mechanism of nacre, Dr Philippe Stemplé for the nanostructure and the mechanical study on nacre, Meriem Lamghari for her nice work with nacre powder injection, Dr Olivier Delattre for the nacre implantation, Professor Anders Meibom for giving me the opportunity to put nacre samples in the NanoSIMS machine, Dr Denis Duplat for his work on nacre degradation and Dr Laurent Bedouet for his important work on nacre molecules. The authors would like to thank Dr. Bernus, veterinary surgeon, for his kind help with the animal surgery. We would particularly thank Mr. G. Mascarel and Prof A. Couté of the Common Service of Electron microscopy of the National Museum of Natural History (Paris, France) for their exceptional scanning electron microscopy artwork of a great part of the studies on nacre.

8. References

Atlan G, *et al.* Interface between bone and nacre implants in sheep. Biomaterials; 20, 1017-1022, 1999.

Atlan G, *et al.* Reconstruction of human maxillary defects with nacre powder: histological evidence for bone regeneration. C R Acad Sci Paris, Life Sci; 320, pp.253-258, 1997.

Bédouet L., *et al.* Low molecular weight molecules as new components of the nacre organic matrix. Comp Biochem Physiol B; 144:532-543, 2006.

Bédouet L., et al. Heterogeneity of proteinase inhibitors in the water-soluble organic matrix from the oyster nacre. Marine Biotechnology; 9, 437-449, 2007.

Bevelander G, Nakahara H. An electron microscope study of the formation of nacreous layer in the shell of certain bivalve molluscs. Calc Tiss Res; 3:84-92, 1969.

Cariolou M, Morse D. Purification and characterization of calcium-binding conchiolin shell peptides from mollusc, *Haliotis rufescens*, as a function of development. J Comp Physiol B; 157:717-729, 1988.

Delattre O., PhD thesis, La nacre de *Pinctada maxima*, biomatériau de substitution et de réparation dans les pertes de substances osseuses et cartilagineuses chez le mouton. Applications potentielles en chirurgie orthopédique, 2000.

Duplat D., *et al.* The in vitro osteoclastic degradation of nacre. Biomaterials; 28, 2155-2162, 2007.

Kono M, *et al.* Molecular mechanism of the nacreous layer formation in *Pinctada maxima*. Biochem Biophys Res Commun; 269:213-218, 2000.

Kulak, A.N., Iddon, P., Li, Y., Armes, S.P., Cölfen, H., Paris, O., Wilson, R.M., Meldrum, F.C., 2007. Continuous structural evolution of calcium carbonate particles: a unifying model of copolymer-mediated crystallization. J. Am. Chem. Soc. 129, 3729-3736.

Lamghari M, et al. Stimulation of bone marrow cells and bone formation by nacre: In vivo and in vitro studies. Bone; 25(2), Suppl.:91S-94S, 1999.

Lamghari M., et al. Bone reactions to nacre injected percutaneously into the vertebrae of sheep. Biomaterials; 22, 555-562, 2001.

Lamghari M., et al., A model for evaluating injectable bone replacements in the vertebrae of sheep: radiological and histological study. Biomaterials, 20, 2107-2114, 1999.

Liao H., et al. Tissue responses to natural aragonite (Margaritifera shell) implants in vivo. Biomaterials; 21, pp.457-468, 2000.

Lopez E, et al. Demonstration of the capacity of nacre to induce bone formation by human osteoblasts maintained in vitro. Tissue Cell; 24, pp.667-679, 1992.

Lopez E., et al. Procédé de preparation de substances actives à partir de la nacre, produits obtenus, utiles notamment comme médicaments, Patent FR9515650, 1995.

Meibom, A., Cuif, J.P., Hillion, F., Constantz, B.R., Juillet-Leclerc, A., Dauphin, Y., Watanabe, T., Dunbar, R.B., 2004. Distribution of magnesium in coral skeleton. Geophys. Res. Lett. 31, L23306.

Meibom, A., Mostefaoui, S., Cuif, J.-P., Dauphin, Y., Houlbreque, F., Dunbar, R.G., Constanz, B., 2007. Biological forcing controls the chemistry of reef-building coral skeleton. Geophys. Res. Lett. 34, L02601.

Pereira-Mouries L, et al. Soluble silk-like organic matrix in the nacreous layer of the bivalve Pinctada maxima. Eur J Biochem; 269:4994-5003, 2002a.

Rousseau M., et al. The water-soluble matrix fraction extracted from the nacre of Pinctada maxima produces earlier mineralization of MC3T3-E1 mouse pre-osteoblasts. Comp Biochem Physiol A; 135:271-278, 2003.

Rousseau M., et al., Low molecular weight molecules of oyster nacre induce mineralization of the MC3T3-E1 cells, J. Biomed. Mater. Res., 85A (2), 487-497, 2007.

Rousseau M., et al., Multi-scale structure of sheet nacre, Biomaterials, 26, 6254-6262, 2005a.

Rousseau M., et al., Restauration of Stratum corneum with nacre lipids, Comp. Biochem. Physiol. Part B, 145, 1-9, 2006.

Shen Y., et al., In vitro osteogenetic activity of pearl. Biomaterials 27, 281-287, 2006.

Silve C., et al., Nacre initiates biomineralization by human osteoblasts maintained in vitro, Calcif. Tissue Int., 51, pp. 363-369, 1992.

Watabe N., J Ultrastructure Research 12: 351-370, 1965.

Weiss IM, et al., Purification and characterization of perlucin and perlustrin, two proteins from the shell of the mollusc Haliotis laevigata. Biochem Biophys Res Commun; 267:17-21, 2000.

Westbroek P, Marin F. A marriage of bone and nacre. Nature; 392:861-862, 1998.

Wohlrab, S., Cölfen, H., Antonietti, M., 2005. Crystalline, porous microspheres made from amino acids by using polymer-induced liquid precursor phases. Angew. Chem. Int. Ed. 44, 4087-4092.

2

Alumina and Zirconia Ceramic for Orthopaedic and Dental Devices

Giulio Maccauro, Pierfrancesco Rossi Iommetti,
Luca Raffaelli and Paolo Francesco Manicone
Catholic University of the Sacred Hearth Rome
Italy

1. Introduction

Ceramic materials are made of an inorganic non-metallic oxide. Usually ceramics are divided into two groups: silicon ceramics and aluminous ceramics. Ceramics are also divided into crystalline and non-crystalline depending on inner molecular organization. Depending on their *in vivo* behaviour, ceramics are classified as bioresorbable, bioreactive or bioinert. Alumina and zirconia are bioinert ceramics; their low reactivity togheter with their good mechanical features (low wear and high stability) led to use them in many biomedical restorative devices. Their most popular application is in arthroprosthetic joints where they have proven to be very effective, that make their use suitable especially in younger, more active patients. Also dental use of these materials was proposed to achieve aesthetic and reliability of dental restorations.

2. Mechanical and chemical features of bioceramics

2.1 Alumina

Corundum known as α-alumina is the alumina ceramic used for biomedical application,. In nature single crystals of this material are known as ruby if containing $Cr2O3$ impurities, or as sapphire if containing titanium impurities which give them blue colour. Al_2O_3 molecule is one of the most stable oxides because of high energetic ionic and covalent bonds between Al and O atoms. These strong bonds (Alumina DG(298K) =1580 KJ/mol) leave the ceramic unaffected by galvanic reactions (absence of corrosion, e.g. absence of ion release from bulk materials and from wear debris). Adverse conditions such as strong acidic or alkaline environment at high temperatures didn't corrupt allumina properties. Under compression allumina showed good resistance but under tensile strength shows its brittleness. at room temperature alumina does not show plastic deformation before fracture (e.g. no yield point in stress-strain curve before fracture), and once started fractures progress very rapidly (low toughness K_{IC}).

Tensile strength of alumina improves with higher density and smaller grain size. A careful selection of raw materials and a strict control of production process are performed by manufacturers to optimize allumina mechanical properties. Introduction of low melting MgO in the ceramic process enhanced mass transport during solid state sintering so that ceramic reached full density at lower temperature. Moreover decreasing grain growth a stronger ceramic was obtained.

Additions of small amounts of Chromia (Cr_2O_3) compensated the reduction of hardness subsequent to the introduction of MgO. CaO content in medical grade alumina devices must be reduced since in wet environment it can compromise its mechanical properties. NaOH impurities in powders obtained by the Bayer process makes allumina unsuitable for the hi-tech biomedical application. Continuous efforts to improve the properties of alumina bioceramics are being made, e.g. by the introduction of high purity raw materials, hot isostatic pressing, proof testing on 100% of manufactured components. The use of hot isostatic pressing (HIP) in bioceramics production minimizes the residual stresses within ceramic pieces and gives ceramics with density close to the theoretical one, improving the strength and reliability of the product. Proof testing of Allumina components consists in the application of internal pressure inducing a stress close to the maximum load bearing capability; when applied to 100% of the parts manufactured, defective products can be eliminated before final inspection. The introduction of laser marking contributed to components traceability due improving the overall quality of the manufacturing process

In 1930 for the first time allumina was used as a biomaterial with the first patent applied by Rock in Germany. Sandhaus in 1965 patented a screw-shaped dental implant made of high alumina powder Degussit AL23. This was the first step in a new era in ceramic engineering. A new dental implant , step-shaped, followed the screw shaped, and was named Tübingen type. But only with the use in orthopaedic purpose in 1970 by Boutin, Allumina was worldwide diffused. He implanted successfully first allumina joints since 1970. Nowadays more than 3 million alumina ball heads have been implanted worldwide. Today, almost 50% of hip arthroplasties performed in Central Europe make use of ceramic ball heads.

2.2 Zirconia

Zirconia, the metal dioxide of zirconium (ZrO_2), was used for a long time as pigment for ceramics; it was identified in 1789 by the German chemist Martin Heinrich Klaproth.

To stabilize zirconium oxide a little amount of non-metallic oxide were added (such as MgO, CaO and Y_2O_3); at a first time magnesia- partially stabilized zirconia (MgPSZ) was the most studied ones, in which a tetragonal phase is present as small acicular precipitates within large cubic grains (Ø40÷50 µm) forming the matrix. But wear properties were badly influenced by this feature; most of the developments were focused on yttria stabilized tetragonal zirconia polycrystal (Y-TZP), a ceramic completely formed by submicron-sized grains, which is today the standard material for clinical applications. Tetragonal grains in Y-TZP ale smaller than 0.5 µm. Tetragonal phase rate retained at room temperature is influenced by: - grains size and on its homogeneous distribution; - the concentration of the yttria stabilizing oxide; - the constraint exerted by the matrix onto grains. The equilibrium among such microstructural parameters influences mechanical features of Y-TZP ceramics. Tetragonal grains can transform in monoclinic, with a 3-4% volume growth of grains: this is the origin of the toughness of the material, e.g. of its ability to dissipate fracture energy. When the pressure on the grains is relieved, i.e. by a crack advancing in the material, the grains near the crack tip shift into monoclinic phase. This gives origin to increased toughness, because the energy of the advancing crack is dissipated at the crack tip in two ways, the first one due to the T-M transformation, the second one due to the need of the crack as it advances to overcome the compression due to the volume expansion of the grains.

In wet environments, over 100°C, tetragonal phase of zirconia ceramics can spontaneously transform into monoclinic. Because this phenomenon starts from the surface of the material, it is possible to report a loss of material density and a reduction in strength and toughness of zirconia. This degradation goes under the name of "ageing" and is due to the progressive spontaneous transformation of the metastable tetragonal phase into the monoclinic phase. Spontaneous T-M transformation in TZP is probably due to the formation of zirconium hydroxides or yttrium hydroxides that promoted phase transition for local stress concentration or variation of the yttrium/zirconium ratio

Swab summarized main steps of TZP ageing in the following way:

The most critical temperature range is 200-300°C.

The effects of ageing are the reduction in strength, toughness and density, and an increase in monoclinic phase content.

Degradation of mechanical properties is due to the T-M transition, taking place with micro and macro-cracking of the material.

T-M transition starts on the surface and progresses into the material bulk.

Reduction in grain size and/or increase in concentration of stabilising oxide reduce the transformation rate.

T-M transformation is enhanced in water or in vapour.

Strength degradation rate is not the same for all TZP ceramics. Swab described that in ten materials tested in presence of water vapour at low temperature, different levels of strength degradation occurred in all the materials but one, where strength remained the same after the treatment. The differences in equilibrium of microstructural parameters like yttria concentration and distribution, grain size, flaw population and distribution in the samples tested caused this variability in ageing behaviour. Strength degradation rate can be controlled by having a high density, small and uniform grain size, a spatial gradient of yttria concentration within grains, introduction of alumina into the matrix. All the above parameters are controlled by the manufacturing process and by the chemical-physic behaviour of the precursors selected for the production of the ceramic. These facts make stability a characteristic of each Y-TZP material and of each manufacturing process.

Hydrothermal treatment has an high risk of phase transition: steam sterilization of zirconia ball heads is not recommended. These process may change the surface finish, reducing wear resistance. Nevertheless, mechanical properties of the material are not altered by these process. Gamma rays or ethylene oxide sterilization are the best choice to manage zirconia biomedical devices. Rare hearth impurities that may be present at part per million (ppm) level within the structure can interact with ionising radiation inducing some changes in colour in ceramic materials. Praseodymium impurities cause a shift to violet of zirconia after irradiation, but the material can return to its ivory colour with heating and putting it under an intense light source; its mechanical properties weren't unaffected by this treatment.

At room temperature Y-TZP ceramic is formed by submicron size grain. during sintering the grains will grow and it is necessary to start from submicron grain size powders and to introduce some sintering aid to limit the phenomenon. The introduction of the stabilizing oxide (yttria Y_2O_3) is a key component in TZP structure at room temperature. hydrothermal stability of the ceramic is enhanced by enriching grain boundaries in yttria: ZrO_2 grains may be yttria coated as in plasma, an alternative to obtain Y-TZP powders by co-precipitation. silica impurities must be avoided because the dissolution of glassy phases at the grain boundaries in wet environment causes the spontaneous transformation of the grains from tetragonal to monoclinic with a loss of mechanical properties. To achieve an equilibrium a higher toughness and hydrothermal stability must be balanced by a lower bending strength

2.3 Zirconia toughened alumina

Zirconia toughened alumina (ZTA) is obtained adding zirconia up to 25% wt into an alumina matrix. This allow to obtain a class of ceramic materials with increased toughness. These materials, developed in the second half of the seventies are featured by toughness (KIC) up to 12 MPam-1/2 and bending strength up to 700 MPa. Alumina matrix exerts a constraint on the metastable tetragonal zirconia particles maintaining them in the tetragonal state. T-M transformation of the zirconia particles give toughness to this ceramic. Because of different elastic modulus between alumina matrix and the zirconia particles cracks are propagated along zirconia crystals inducing their T-M phase transformation thus dissipating the crack energy. Microcracking of the matrix due to the expansion of the dispersed particles is a further dissipative effect. To ensure the better mechanical performances to this material is mandatory to control the high density of the matrix and the optimisation of the microstructure of the zirconia particles. In this way the maximum amount of metastable phase is retained assuring the transformation of the maximum volume. When hardness is of paramount importance ZTA have some drawbacks: zirconia into the hard alumina matrix results in a decrease in hardness of the ceramic Extensive research has been focussed on ZTA in France and in Italy on ceramics containing up to 80% zirconia, without leading to clinical applications. Allumina can also be toughened by addition of whiskers; but concerns about carcinogenicity of whiskers, and limits in adhesion of the whiskers to the matrix decreased the interest for the biomedical applications of these materials. Elongated grains (platelets), acting as whiskers, can be nucleated within the structure of a ZTA ceramic. This can be obtained by adding e.g. strontium oxide (SrO) to ZTA obtaining SrAl12O19platelets by in situ solid state reaction during sintering. Chromia (Cr2O3), introduced to save the alumina hardness and of Yttria (Y2O3) that acts as stabilizer of the T phase of zirconia in ZTA, leads to a material known as ZPTA(Zirconia Platelet Toughened Alumina) The resulting mechanical properties are very interesting, as wear rates were very low in the laboratory tests, even lower than the ones of alumina and zirconia both on hip and knee simulator studies

ZPTA is a great innovation in ceramic for biomedical devices. Mechanical properties of this new ceramic, allow to develop many innovative ceramic devices.

Property	Unit	Allumina	Y-TZP	ZTA	ZPTA
Density	g/cm^2	3.98	6.08	5.00	4.36
Average grain size	μm	≤1.8	0.3÷0.5	-	-
Bending strength	MPa	>550	1200	900	1150
Compression strenght	MPa	5000	2200	2900	4700
Young modulus	GPa	380	200	285	350
Fracture toughness K_{IC}	$M_{pam}^{-1/2}$	4-5	9	6.9	8.5
Microhardness	HV	2200	1000-1300	1500	1975

Table 1. Selected Properties of load bearing bioceramics for medical devices

3. Biocompatibility

Biocompatibility has been defined as "the ability of a material to perform with an appropriate host response in a specific application". Reaction of bone, soft collagenous tissues and blood are involved in the host response to ceramic implants. Interfacial reaction between these materials and body tissues both *in vitro* and *in vivo* must be considered evaluating biocompatibility of bioinert ceramics. Low rate of tissue reactions towards Alumina are the reason because it is often considered reference in testing orthopaedic ceramic biomaterials. The first experimental data of dense ceramics (ZrO2) *in vivo* biocompatibility in orthopaedic surgery were published 1969 by Helmer and Driskell while the first clinical cases on alumina were described later by Boutin shortly followed by Griss. *In vitro* biocompatibility evaluation of Alumina and Zirconia were performed later than their clinical use. As biocompatibility tests often are reporting the comparison of alumina and zirconia biocompatibility, in the following the results are reviewed in the same manner.

3.1 In vitro tests

Ceramic materials in different physical forms (powders and dense ceramics) were used to perform *in vitro* tests on cell cultures. Absence of acute toxic effects of ceramic in powder and disk form on the different cell lines used in tests both towards allumina both toward zirconia was reported by many studies. *In vitro* assays are influenced by material characteristics, such as the physical form, reactive surface, chemical composition, impurity content etc, as well as by the cell conditions during the tests. Alumina and zirconia disks with 30% of porosity allow adhesion and spreading of 3T3 fibroblasts as observed using SEM. HUVEC and 3T3 fibroblasts osteoblast didn't show any toxic reaction toward Al2O3 or ZrO2 samples (MTT test on cells direct in contact with ceramic particles); the same effects were also observed on ceramic extracts cocultured with fibroblasts. Li, et al demonstrated that powders were more toxic than dense ceramics, using direct contact tests and MTT test with human oral fibroblasts. Ceramic powders can induce apoptosis in macrophages depending on materials concentration as observed by Catelas. Mebouta, et al reported for the first time a different toxic effect between alumina and zirconia: in particular a higher cytotoxicity of alumina particles in comparison to the zirconia ones was measured as human monocytes differentiation; this is probably due to the higher reactive surface of the alumina particles, that were significantly smaller than the zirconia ones

Degidi compared soft tissues reactions to ZrO_2 and titanium; he reported that inflammatory infiltrate, microvessel density and vascular endothelial growth factor expression appeared higher around titanium samples than around ZrO_2 ones. Moreover cellular proliferation on zirconia surface is higher than on titanium ones. Furthermore Warashima reported less proinflammatory mediators(IL-1β, IL-6 and TNF–α) generated by ZrO_2 than titanium or polyethylene.

3.2 In vivo tests

Different physical forms and in different sites of implantation were evaluated in order to analyzing systemic toxicity, adverse reactions of ceramics in soft tissue and/or bone The work of Helmer and Driskell already cited is the first report of implant in bone of zirconia. Pellets were implanted into the monkey's femur, the Authors observed an apparent bone ingrowth without any adverse tissue reaction. Hulbert, et al implantated of porous and non porous disks and tubes in the paraspinal muscles of different ceramics Authors observed ingrowth depending on porous size, and no signs of systemic toxicity. After subcutis, intramuscular or intraperitoneal and intraarticular introduction of alumina and zirconia powders in rats and/or mice mant authors reported the absence of acute systemic adverse tissue reactions to ceramics; similar results were reported after implantation of bars or pins to paraspinal muscles of rabbits or rats and after insertion in bone. bone ceramic interface showed connective tissue presence, progressively transformed in bone direct contact with ceramic. Bortz reported adverse tissue reaction: fibrous tissue in the lumen of zirconia cylinders implanted in dogs and rabbits trachea, and an inflammatory reaction against ceramic powders inserted on PMMA grooves implanted in rabbits femur. In any case this inflammatory reaction was lower than the one observed against CoCr and UHMWPE.

3.3 Carcinogenicity
Griss, et al. in 1973 reported that Alumina and zirconia powders did not induce tumours. They analyzed the long term in vivo reactions to ceramics.

Ames test, and carcinogenicic or mutagenic tests used to study zirconia dishes confirmed that this bioceramic did not elicit any mutagenic effect in vitro. Moreover zirconia radioactivity and its possible carcinogenic effect was also evaluated: radioactivity of the powder is depending on the source of ores used in the production of the chemical precursor of the zirconia powders. Only Ryu RK, et al reported a possible carcinogenic effect of ceramic. They observed association between ceramic and soft tissue sarcoma. Some recent studies have been performed about carcinogenicity of Zirconia Toughened Alumina. Maccauro et al. showed that ZPTA as well as Alumina and Zirconia ceramics did not elicit any in vitro carcinogenic effects; the same group are going to demonstrate the possible carcinogenic in vivo effects of ZPTA.

4. Biomedical applications of zirconia
Several comprehensive reviews on the clinical outcomes of ceramic ball heads for orthopaedical devices are aviable. Jenny, Caton, Oonishi, Hamadouche, demonstrate the favourable behaviour of ceramic biomaterials in reducing the wear of arthroprostheses joints.

THR ball heads
THR acetabular inlays
THR condyles
Finger joints
Spinal spacers
Humeral epiphysis
Hip endoprostheses

Table 2. Orthopaedic medical devices made of bioinert ceramics

Clinical trials demonstrated that ceramic-on-ceramic coupling decreased significantly the amount of wear debris (Boeler,). Nevertheless Wroblesky demonstrated that ceramic in couple with new generation polyethylene may constitute a significant evolution in arthroplasty. This makes ceramics in joints suitable especially in younger patients. The matching of surface roughness, roundness and linearity in the coupling of ceramic tapers with the metallic trunnion plays a relevant role on stresses distribution and intensity, depending also on cone angle, extent of the contact, friction coefficient among the two surfaces Mismatch in female- to-male taper, e.g. due to the many angles in clinical use, roundness, roughness or linearity errors in the taper, are among the most likely "technologic" initiators of ceramic ball head failures. It must be remarked that the mechanical behaviour of ceramic ball heads once installed on the metallic taper depend not only on the ceramic but also on material and design of the taper. Besides the "technologic" failure initiators, several other precautions are necessary when using ceramic ball heads: avoid third body interposition to the ceramic metal, or ceramic/ceramic interface during surgery (e.g.blood clots, bone chips, PMMA cement debris); avoid use of metallic mallets when positioning ball heads on metallic taper (or of alumina inlay into the metal back): use plastic tools provided by the manufacturer or gently push rotate by hand; avoid thermal shocks to ball heads (e.g. dip the ceramic in saline to cool it after autoclave sterilization); avoid application of new ceramic ball heads onto stems damaged during revision surgery. A third important aspect to achieve good arthroprostheses results is surgical technique: both perfect THR component adaptation and orientation, together with soft tissue tension are required. Special care must be taken with orientation, as edge loading of the socket and impingement on components depend on this parameter.

In the past zirconia was highly used in orthopedics; about 900000 zirconia ball heads have been implanted in total hip arthroplasties, even if a debate arose regarding the potential radioactivity and carcinogenicity of zirconia source. But, after the observation of some ball head fractures, zirconia has no longer been used for total hip arthroplasties.

Zirconium oxide is also used as a dental restorative material. Inlays, onlays, single crowns, fixed partial dentures, can be realized using a ZrO_2 core. Moreover, also implant abutment and osteointegrated implant for tooth replacement are available in zirconia.

Realization of dental products requires a preventive project and successive manufacturing in order to satisfy clinical requirement. But, not only individualization is needed: accuracy is absolutely mandatory. Misfits greater than 50 μm are considered unacceptable for dental restorations. Mechanical resistance must be also considered. Frameworks with minimal thickness, often less than 1mm, must be able to sustain chewing stresses. Masticatory load on posterior teeth range from 50N to 250N, while parafunctional behavior such as clenching and bruxism can create loads about 500 and 800N. Zirconia frameworks can bear load between 800 and 3450N. These values are compatible with restorations on posterior teeth if parafunctional loads are not present and a correct framework design is performed .

In order to avoid misfit due to shrinkage during sintering, it is possible to obtain zirconia frameworks by milling full-sintered ZrO_2 samples. This technique is not influenced by sintering problems because zirconia is already sintered, but, anyway, it is influenced by operator accuracy in probe use. CAD/CAM technique is the ultimate opportunity in managing zirconia dental devices production. CAD/CAM is acronym of Computer Aided Design and of Computer Assisted Manufacturing. This system is composed by a digitizing machine to collect information about teeth position and shape, appropriate software for design zirconia restoration and a computer assisted milling machine that cut from a zirconia sample the desidered framework. This technique reduces human influences allowing obtaining greater accuracy in zirconia core production.

Fully sintered zirconia blocks are very difficult to be grinded. Milling procedures are very slow and requires very effective burs to perform cut in the optimal way. Dimensional stability is granted because there aren't any procedures that can influence volume of framework after milling. On the other hand, grinding reduces significatively toughness of zirconia. This can be due to surfacial stresses during milling. Crystals were induced to transform from tetragonal into monocline reducing T-M phases ratio and consequently toughness. Lutardth measured flexural strength and fracture toughness of zirconia before and after grinding and concluded that mechanical resistance was reduced of about 50% after machining. Also Kosmac studying surface grinding effects on ZrO_2 confirmed these results.

Machining partially sintered zirconia (or green zirconia) presents, on the other hand, different problematics. Green zirconia has a very soft consistency resulting very easy to be milled. Grinding procedures are easy, faster and cheaper. But, after grinding, frameworks must be sintered. This procedure presents some technical problems that require accurate managing to grant a reliable outcome. During Sintering time (about 11 hours) an accurate control of temperature and pressure, especially during cooling phase, is needed to obtain the correct T-M crystals ratio. Moreover, sintering lead to a 20% volumetric shrinkage that must be foresight in advance during designing and milling. For these reasons use of green zirconia results more difficult and expensive: complex designing software and sintering machine are required to obtain accuracy and correct crystal composition. On the other side, if procedures are preformed correctly mechanical resistance results greater than ZrO_2 frameworks milled after sintering]. Moreover sinterization after grinding allows technician also to pigment frameworks helping achieving a satisfying aesthetical outcome.

Ceramic restorations allow an aesthetical outcome more similar to teeth than conventional metal-ceramic ones. Also gingival aesthetic is improved by colour of restoration similar to teeth, that is, together with mucosal thickness the basic parameter for an optimal soft tissue colour outcome. Toughness and colour similar to teeth of zirconia, lead to use this material for different purpose. Zirconiun oxide is used as a reinforce for endocanalar fiber-glass post. Also orthodontic brackets were proposed in ZrO_2. But the most interesting application of this material is nowadays for fixed partial dentures. Single crowns and 3-5 units FPD are described and studied in literature. The continuative search for an optimal metal-free material for prosthetical use found in zirconia an answer for many problems still not solved with other ceramic restorations. Also small dental restorations, like inlays and onlays were proposed with this material. Implanto-prothesical components, such as implant abutment are available with zirconia. Osteointegrated implants for tooth replacement are proposed by some manufacturers, but at the present time there aren't enough studies about behaviour of zirconia implants.

Zirconia restorations have found their indications for FPDs supported by teeth or implants. Single tooth restorations are possible on both anterior and posterior elements because of the mechanical reliability of this material. Mechanical resistance of zirconia FPD was studied on single tooth restorations and on partial dentures. Luthy asserted that Zirconia core could fracture with a 706N load Tinshert reported a fracture loading for ZrO_2 over 2000N, Sundh measured fracture load between 2700-4100N. Zirconia restorations can reach best results as fracture resistance if compared with alumina or lithium disilicate ceramic restorations.

Ageing of Zirconia can have detrimental effects on its mechanical properties. To accelerate this process mechanical stresses and wetness exposure are critical. Ageing on zirconium oxide used for oral rehabilitation is not completely understood. However, an in vitro simulation reported that, although ageing the loss of mechanical features does not influence resistance under clinical acceptable values. Further evaluations are needed because zirconia behavior in long time period is not yet investigated.

On these basis a new family of ceramic material that would complement alumina ceramic where needed. It had to posses the highest possible toughness, the smallest matrix grain size all leading towards improved mechanical reliability but this had to be accomplished without sacrificing the wear resistance and chemical stability of current day alumina ceramics. Alumina Matrix Composites were selected as the best new family of ceramics to provide the foundation for an expanded use of ceramics in orthopaedics. The main characteristics of this Alumina Matrix Composite are its two toughening mechanisms. One is given by in-situ grown platelets which have a hexagonal structure and are homogeneously dispersed in the microstructure. Their task is to deflect any sub-critical cracks created during the lifetime of the ceramic and to give the entire composite stability. The other important characteristic is related to the addition of 17 vol.-% zirconia nano-particles that are dispersed homogeneously and individually in the alumina matrix. This increases strength and toughness of the material to levels equal and in some cases above those seen in pure zirconia. Here, the effect of tetragonal to monoclinic phase transformation is used as a toughening mechanism. In the case of micro-crack initiation the local stress triggers phase transformation at an individual zirconia grain which acts then as an obstacle to further crack propagation. It is a desired behaviour which uses the volume expansion in an attempt to prevent further crack propagation. These two well known effects in material science, crack deflection and transformation toughening give Alumina Matrix Composite a unique strength and toughness unattained by any other ceramic material used in a structural application in the human body.

5. References

Carinci, F.; Pezzetti, F.; Volinia, S.; Francioso, F.; Arcelli, D.; Farina, E. & Piattelli, A. (2004) Zirconium Oxide: Analysis Of Mg63 Osteoblast-Like Cell Response By Means Of A Microarray Technology. *Biomaterials* Vol.25 p.215.

Christel, P.; Meunier, A.; Dorlot, J-M. (1988) Biomechanical Compatibility And Design Of Ceramic Implants For Orthopedic Surgery, In: *Bioceramics: Material Characteristics Versus In Vivo Behavior.* Ducheyne, P. & Lemons, J. E. Annals Of New York Academy Of Sciences,Vol. 523 pp. 234-256.

Covacci, V.; Bruzzese, N. & Maccauro, G. (1999) In Vitro evaluation Of The Mutagenic and carcinogenic power of high purity zirconia ceramics. *Biomaterials* Vol. 20 pp.371-376

Dion, I.; Bordenave, L. & Lefebvre, F. (1994) Physico-Chemistry And Cytotoxicity Of Ceramics, Part 2 *J. Mater. Sci. Mater. Med.* Vol. 5 pp. 18-24

Fenollosa, J.; Seminario, P. & Montijano C. (2000) Ceramic Hip Prostheses In Young Patients - A Retrospective Study. *Clin Orthop;* Vol.379 pp. 55-67

Heimke, G.; Leyen, S. & Willmann, G. (2002) Knee Arthroplasty: Recently Developed Ceramics Offer New Solutions. *Biomaterials* Vol.23 pp.1539-51

Helmer, J.D. & Driskell, T.D, (1969) *Research On Bioceramics.* Symp. On Use Of Ceramics As Surgical Implants. South Carolina (Usa): Clemson University.

Kingery, W.D.; Bowen, H.K. & Uhlmann, D.R. (1960) *Introduction To Ceramics*. John Wiley & Sons Publ. USA

Maccauro, G.; Cittadini, A.; Magnani, G.; Sangiorgi, S.; Muratori, F.; Manicone, P.F.; Rossi Iommetti, P.; Marotta, D.; Chierichini, A.; Raffaelli, L. & Sgambato, A. (2010) In Vivo Characterization Of Zirconia Toughened Alumina Material: A Comparative Animal Study. *Int J Immunopathol Pharmacol*. Vol. 23 pp.841-6.

Maccauro, G.; Piconi, C. & Burger, W. (2004) Fracture Of A Y-Tzpceramic Femoral Head. Analysis Of A Fault. *J Bone Joint Surg Br* Vol. 86 pp.1192-6.

Maccauro, G.; Bianchino, G.; Sangiorgi, S.; Magnani, G.; Marotta, D.; Manicone, P.F.; Raffaelli, L.; Rossi Iommetti, P.; Stewart, A.; Cittadini, A. & Sgambato, A. (2009) Development Of A New Zirconia-Toughened Alumina: Promising Mechanical Properties And Absence Of In Vitro Carcinogenicity. *Int J Immunopathol Pharmacol*. Vol. 22 pp.773-9.

Manicone, P.F.; Rossi Iommetti, P. & Raffaelli, L. (2007) An Overview Of Zirconia Ceramics: Basic Properties And Clinical Applications. *J Dent*. Vol. 35 pp. 819-26.

Mendes, D.G.; Said, M. & Zukerman, V. (2000) Ten Rules Of Technique For Ceramic Bearing Surfaces In Total Hip Arthroplasty. In: *Bioceramics In Total Hip Joint Replacement*. Willman, G. & Zweymüller, K. Stuttgart: Thieme Publ pp.9-11.

Oonishi, H.; Amino, H.; Ueno, M. & Yunoki, K. (1999) Concepts And Design With Ceramics For Total Hip And Knee Replacement. In: *Reliability And Long Term Results Of Ceramics In Orthopedics*. Sedel, L. Willmann, G. Stuttgart, Germany: Thieme Publ pp. 7-28.

Piconi, C.; Burger,W. & Richter, H.G. (1998) Y-Tzp Ceramics For Artificial Joint Replacements. *Biomaterials* Vol. 19 pp.1489-1494

Piconi, C. & Maccauro, G. (1999) Zirconia As A Ceramic Biomaterial. *Biomaterials* Vol. 20 pp. 1-25.

Raffaelli, L.; Rossi Iommetti, P.; Piccioni, E.; Toesca, A.; Serini, S., Resci, F.; Missori, M.; De Spirito, M.; Manicone, P.F. & Calviello, G. (2008) Growth, Viability, Adhesion Potential, And Fibronectin Expression In Fibroblasts Cultured On Zirconia Or Feldspatic Ceramics In Vitro. *J Biomed Mater Res A*. Vol.86 pp.959-68.

Rieger, W. (2001) Ceramics In Orthopaedics – 30 Years Of Evolution And Experience. In: *World Tribology Forum In Arthroplasty* Rieker , C.; Oberholtzer, S. & Wyss, U. Hans Huber Publ, Bern, Ch, pp.309-318.

Salomoni, A.; Tucci, A.; Esposito, L. & Stamenkovich, I. (1994) Forming And Sintering Of Multiphase Bioceramics, *J Mater Sci Mater Med* Vol. 5 pp. 651-653.

Sato, T. & Shimada, M. (1985) Control Of The Tetragonal-To-Monoclinic Phase Transformation Of Yttria Partially Stabilized Zirconia In Hot Water. *J Mater Sci*. Vol. 20 pp. 3899-992.

Swab, J.J.(1991) Low Temperature Degradation Of Y-Tzp Materials. *J Mater Sci* Vol.26 pp.6706-14.

Vigolo, P.; Fonzi, F.; Majzoub, Z. & Cordioli, G. (2006) An In Vitro Evaluation Of Titanium, Zirconia, And Alumina Procera Abutments With Hexagonal Connection. *International Journal Of Oral Maxillofacial Implants* Vol. 21 pp.575-80.

3

Polysaccharides as Excipients for Ocular Topical Formulations

Ylenia Zambito and Giacomo Di Colo
University of Pisa
Italy

1. Introduction

The topical treatment of extraocular or intraocular diseases, especially by eye drops, is the best accepted by patients. Treatment with eye drops, however, poses the issue of a poor drug bioavailability because the precorneal area, i. e., the site of drug action/absorption, is rapidly cleared of drugs by protective mechanisms of the eye, such as blinking, basal and reflex tearing, and nasolachrymal drainage. This implies the need of frequent instillations, and hence, the risk of side effects. Increasing ocular bioavailability remains a stimulating challenge for the formulators of topical systems. An approach to the task has been the reduction of drainage rate by increasing the viscosity of liquid preparations (Lee & Robinson, 1986) or resorting to mucoadhesive polymers (Hui & Robinson, 1985). The ability of a polymer to improve the ocular bioavailability of drugs by adhering to the ocular surface and binding the drug to it is a more promising property than the polymer viscosifying power (Di Colo et al., 2009), so far as fluid solutions are better tolerated than viscous ones (Winfield et al., 1990). The ocular retention of drugs administered by eye drops is also potentially improved by colloidal drug carriers, such as liposomes, submicron emulsions, nanoparticles and nanocapsules. The drugs are incorporated into these submicron particles which can be internalized into the corneal and/or conjunctival cells of the ocular epithelium (Alonso & Sanchez, 2009; Nagarwal et al., 2009). Prolonging the residence of drugs in the precorneal area serves either the extra- or intraocular therapy. In those cases where a well tolerated topical treatment is desired to implement an intraocular therapy, the ocular formulation can be made to contain an effective, biocompatible, non-irritant polymeric corneal permeability enhancer.

A description of the structures of the eye that come into contact with topical drug delivery systems has been reported recently (Ludwig, 2005). A summary is given in the next section, followed by an outline of the routes of intraocular drug penetration. Polysaccharides such as chitosan, xyloglucan, arabinogalactan, cellulose derivatives (methylcellulose, hydroxyethylcellulose, hydroxypropylmethylcellulose, sodium carboxymethylcellulose), hyaluronic acid, alginic acid, gellan gum, have been studied extensively as excipients for ocular formulations. In the present survey of the literature the relevant properties of polysaccharides will be presented and discussed, with emphasis on the functions of polymers in the ophthalmic formulations where they have been used. Only polysaccharides the ocular tolerability of which has been ascertained will be dealt with.

2. Anatomy and physiology of the eye

2.1 Structure of the ocular globe

The human eye is schematically shown in Fig.1. The wall of the eyeball consists of an outer coat (sclera and cornea), a middle layer (uveal coat) and an inner coat (retina). The eyelids spread the tear fluid over the eye. The rate of shear during blinking (about 20,000 s^{-1}) influences the rheological properties of instilled ophthalmic formulations and hence, drug bioavailability.

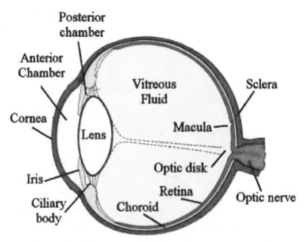

Fig. 1. Schematic view of the human eye

The cornea is a clear, avascular tissue composed of five layers: epithelium, Bowman's layer, stroma, Descemet's membrane and endothelium. The epithelium consists of 5-6 cell layers. The corneal epithelium is little permeable due to the presence of tight junctions connecting cells. High extracellular and low intracellular calcium levels are required for the low permeability of tight junctions. On the surface of the epithelium flattened cells are found microvilli which enhance the stability of the tear film.

The conjunctiva is a clear membrane, lining the inner surface of the eyelids, adjoining the corneal epithelium. The conjunctiva is vascularized and moistened by the tear film. Its epithelium is composed of 5-7 cell layers. The cells are connected by tight junctions, therefore the conjunctiva is impermeable to molecules larger than 20,000 Da, whereas the cornea is permeable up to 5000 Da.

A volume of about 2-3 µL of mucus is secreted daily whereby foreign particles and bacteria are entrapped and swept by blinking to the drainage system for discharge. The turnover of the mucous layer (15-20 h) is much slower than that of the tear fluid.

2.2 Nasolachrymal drainage system

The tear fluid is spread on the ocular surface by blinking and collected by the canaliculi, the lachrymal sac and the nasolachrymal duct which opens into the inferior nasal passage. The tear fluid produced by the lachrymal gland is composed of the following:

1. Basic tearing (0.5-2.2 µL/min), needed to maintain a tear film on the corneal surface, corresponding to a turnover rate of 16%/min while awake.

2. Reflex tearing, caused by such stimuli as emotional, chemical or mechanical ones, cold temperature, light, which can raise lachrymation up to 300 µL/min, thus clearing drugs away.

Undesired drug absorption through nasal mucosa can occur during drainage.

2.3 Tear film

The precorneal tear film (thickness 3-10 µm; volume ~10 µL) is composed of the following layers:

The superficial lipid layer (thickness 100 nm), which spreads over the aqueous layer during eye opening. It contains such lipids as triglycerides, phospholipids, sterols, sterol esters, fatty acids, and helps tear fluid to maintain its normal osmolality by limiting evaporation.

The aqueous layer, which contains inorganic salts, glucose, urea, retinol, ascorbic acid, proteins, lipocalins, immunoglobulins, lysozyme, lactoferrin and glycoproteins. The tear fluid has an osmolality of 310-350 mOsm/kg and a mean pH of 7.4. Its buffering ability is determined by bicarbonate ions, proteins and mucins. The viscosity is ~3 mPas, with a non-Newtonian rheological behaviour. The surface tension is ~44 mN/m.

The mucus layer, which forms a gel with viscoelastic properties. Mucins improve the spreading of the tear film and enhance its stability and cohesion. The mucus gel entraps bacteria, cell debris and foreign bodies forming bundles of thick fibers, which are conveyed by blinking to the inner canthus and expelled onto the skin. Mucus, which is charged negatively, can bind positively charged substances.

More recent studies have somewhat revised the above "three layers theory", proposing a 40-µm thick film essentially made of mucus with an external lipid layer, but without a distinct thin aqueous layer. Mucus is made of glycoproteins (mucins), proteins, lipids, electrolytes, enzymes, mucopolysaccharides and water. As the polysaccharide side chains of mucins usually terminate in sialic acid (pKa 2.6) the mucins are negatively charged in physiological conditions. A combination of cross-linking via disulfide bridges and hydrophobic bonds and also through entanglements of randomly coiled macromolecules determines the tertiary structure of mucin.

Various factors influence the mucoadhesion of polymeric ocular delivery systems, linked to the composition, physicochemical properties and structure of the tear film. The polymer must come into intimate contact with the mucus layer. The polymer chains must be mobile and flexible enough to interpenetrate into the mucus to a depth sufficient to create a strong entangled network with mucin. The polymer and mucin should interact by hydrogen bonding, electrostatic and hydrophobic interactions which depend on the ionic strength and pH of the applied vehicle. Decreasing the pH enhances the mucoadhesion of polymers containing carboxyls because these groups preferentially interact with mucins in the unionized form via hydrogen bonding. On the other hand, the repulsion of carboxylate anions which form at higher pH values causes extension of polymer chains and decrease in density of such chains, which result in enhancement of chain mobility, interdiffusion and physical entanglement. Depending on the pKa of the functional groups of the polymer, either hydrogen bonding or entanglement prevails. Then, the precorneal residence of mucoadhesive polymers varies from a few hours to one day.

The tear film is not quite stable. In the short interval between blinks it ruptures with formation of dry spots on the cornea, which induces blinking and spreading of a new tear film. The breakup time of the tear film depends on dispersion forces, interfacial tension and viscous resistance of the mucus layer. These factors should be taken into account when developing mucoadhesive pharmaceutical systems.

3. Routes of intraocular penetration

Drugs penetrate across the epithelium into the eye *via* the transcellular (lipophilic drugs) or paracellular (hydrophilic drugs) route (Nanjawade et al., 2007). The main mechanism of transepithelial either transcellular or paracellular intraocular penetration of topically applied drugs is passive diffusion along their concentration gradients. The cornea provides the rate-limiting resistance against penetration of hydrophilic drugs, whereas partitioning from the epithelium into the hydrophilic stroma is the rate-liming factor for lipophilic drugs (Nanjawade et al., 2007). Permeation of ionizable drugs depends on the chemical equilibrium between ionized and unionized species in the tear fluid. The unionized species usually penetrates more easily than the ionized species. In the latter instance the type of charge affects transcorneal penetration. Indeed, the corneal epithelium is negatively charged, in physiological conditions (pH 7.4), hence cationic drugs permeate more easily than anionic ones (Nanjawade et al., 2007). In addition to the corneal route, topically applied drugs may be absorbed *via* non-corneal routes. Hydrophilic and large molecules, showing poor corneal permeability, can penetrate across the bulbar conjunctiva and underlying sclera into the uveal tract and vitreous humor (Nanjawade et al., 2007). Tight junctions of the superficial conjunctival epithelium are wider than those in the cornea, therefore, the conjunctival permeability of hydrophilic drugs is generally significantly higher than their corneal permeability (Nanjawade et al., 2007). However, the conjunctiva is vascularized, therefore it is generally considered a site for systemic, hence, unproductive absorption.

4. Polysaccharides in ocular formulations

The use of natural polysaccharides in ocular formulations is attractive as these products are economical, readily available, non-toxic, potentially biodegradable, generally biocompatible and capable of chemical modifications. Such modifications have recently led to derivatives with improved biopharmaceutical performances, e.g. rhelogical behavior, mucoadhesivity, ability to enhance corneal permeability, easy solubilization.

4.1 Chitosan and its derivatives

Chitosan (Ch) is a linear polysaccharide composed of randomly distributed β-(1-4)-linked D-glucosamine (deacetylated unit) and N-acetyl-D-glucosamine (acetylated unit) (Fig.2).

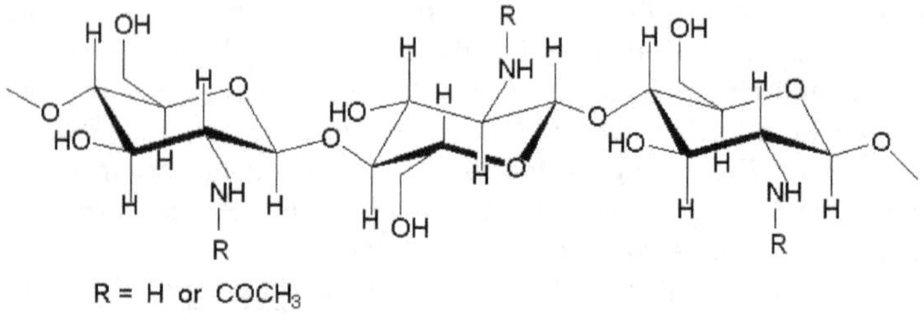

R = H or COCH₃

Fig. 2. Chitosan structure

Ch is produced commercially by deacetylation of chitin, which is the structural element in the exoskeleton of crustaceans (crabs, shrimps, etc.) and cell walls of fungi. The degree of deacetylation in commercial Chs is in the range of 60-100 %. The tremendous potential of Ch in the pharmaceutical area is illustrated in several review articles (Alonso & Sanchez, 2009; Nagarwal et al., 2009; Dodane & Vilivalam, 1998; Felt et al., 1998; Paul & Sharma, 2000; Singla & Chawla, 2001, Di Colo et al., 2008). These point out that Ch is biodegradable, has low toxicity and good ocular tolerability, exhibits bioadhesion and permeability-enhancing properties and also physico-chemical characteristics that make it suitable for the design of ocular drug delivery vehicles. The Ch repeating unit bears a primary amino group bestowing a reactivity on the polymer that allows its transformation into derivatives of interest as biocompatible and bioactive excipients for ophthalmic drug delivery systems (Di Colo et al., 2004a; Zambito et al., 2006a; Zambito et al., 2007). Over the last decade efforts have been made to put into evidence the ability of Ch to safely promote intraocular drug penetration by enhancing corneal permeability, and to synthesize Ch derivatives with improved such bioactivity. The significant results of these efforts will be discussed in the next subsections.

4.1.1 Eye drops

The main role of Ch in ocular solutions was to improve ocular drug bioavailability by prolonging drug residence in precorneal area in virtue of polymer mucoadhesivity and effect on viscosity (Alonso & Sancez, 2009). More recently Ch solutions have been investigated for their ability to improve intraocular drug bioavailability, following topical administration, by enhancing corneal permeability. Although the ability of Ch to permeabilize nasal mucosa and intestinal epithelium has been known since 1994 (Illum et al., 1994; Artursson et al., 1994) the permeabilizing effect of this polymer applied via eye drops on the cornea was demonstrated for the first time ten years later (Di Colo et al., 2004b). Unlike the intestinal and nasal epithelia, both made of a single layer of cells joined by tight junctions, the cornea is a stratified epithelium the cells of which also are joined by tight junctions. Therefore, the ability of Ch to enhance the permeability of intestinal and nasal epithelia would not necessarily imply a similar effectiveness on the corneal epithelium.

Instead of using the excised cornea, as in previous studies of the enhancement of corneal permeability to drugs, the permeabilizing effect of Ch was investigated *in vivo* (Di Colo et al., 2004b) . To this purpose, ofloxacin was instilled into the eyes of albino rabbits via isoviscous eye drops containing Ch hydrochloride (Ch-HCl), or N-carboxymethyl Ch (CMCh) (Muzzarelli et al., 1982), or poly(vinyl alcohol) (PVA), and the resulting pharmacokinetics in tear fluid and aqueous humor were determined (Di Colo et al., 2004b). CMCh, a polyanion at the physiological pH of the tear fluid (Fig.3), was tested to ascertain the relevance of the polycationic nature of Ch-HCl and because CMCh was claimed to behave as an intestinal absorption enhancer (Thanou et al., 2001).

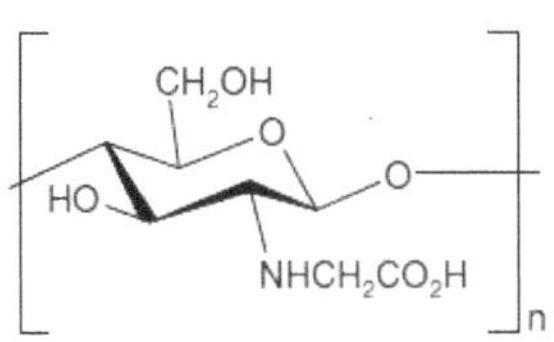

Fig. 3. Structure of N-carboxymethyl chitosan.

Only small differences among the rates of drug disappearance from tear fluid were measured for the three solutions, reflecting the small differences among the respective viscosity values. Such values were high enough to ensure the presence of drug in tear fluid at measurable concentrations after about 1.5 h of instillation. Although the time of drug residence in the precorneal area was almost the same for the three solutions containing Ch-HCl, CMCh or PVA, the respective pharmacokinetic data for the aqueous humor were neatly distinct. This was taken as a sign of different effects of polymers on the corneal permeability. PVA produced an increase of t_{max} with respect to the reference Exocin®, which was ascribed to the increased viscosity of the PVA solution with respect to the commercial eye drops, causing a reduction of tear fluid drainage, and hence, of the precorneal elimination rate. This polymer, however, produced no permeabilization of the cornea, indeed, there was no significant increase of neither concentration peak (C_{max}) nor bioavailability (AUC) in the aqueous. On the other hand, Ch-HCl produced C_{max} and AUC values remarkably higher than the respective values for the reference, at the same t_{max}. Then unlike the case of PVA, with Ch-HCl the increased viscosity of the solution was found to cause no prolongation of t_{max}. These pharmacokinetic data were taken as indicative of an increase of the corneal absorption rate constant, due to an enhancement of corneal permeability by Ch-HCl.

More recently the effects of Ch and other non-polymeric permeabilizers on the permeation of acyclovir across excised rabbit cornea have been studied (Majumdar et al., 2008). Ch was solubilized at the concentration of 0.1 or 0.2% by 2% acetic acid added to Dulbecco's phosphate buffered saline. The apparent corneal permeability (P_{app}) was determined by normalizing the permeant flux to the permeant concentration in the donor. In the presence of 0.2 and 0.1% Ch the transcorneal acyclovir permeability was enhanced almost 5.8-fold (7.61×10^{-6} cm/s) and 3.1-fold (4.1×10^{-6} cm/s), respectively, over that of acyclovir alone (1.32×10^{-6} cm/s).

The use of unmodified Ch in eye drops as a cornea-permeabilizing agent is not quite rational because of problems with its solubility in tear fluid at the physiological pH of 7.4. Although Ch-HCl is in the dissolved, and hence, bioactive state just as applied, yet its permeability-enhancing effect is presumably temporary. Indeed Ch requires pH ≤ 5 for dissolution, hence it is expected to precipitate as the free base some time after instillation, as soon as the physiological pH of the tear fluid is restored. For this reason the effect of CMCh, a polyanionic Ch derivative soluble at this pH (Fig.3), on transcorneal drug absorption is of particular relevance. Although CMCh was found to behave as an intestinal absorption enhancer (Thanou et al., 2001) , it failed to significantly enhance corneal permeability, in fact, the drug levels produced in the aqueous by CMCh never exceeded the C_{max} for the reference Exocin® (Di Colo et al., 2004b). Then it appears that the polycationic nature of Ch is essential to its permeabilizing effect on the cornea. Interestingly, however, the drug concentration in the aqueous vs. time data relative to CMCh pointed to zero-order transcorneal absorption kinetics. This observation prompted the hypothesis that CMCh might mediate a pseudo-steady-state transcorneal transport, via polymer interactions with the drug and polymer adhesion to the corneal mucus (Di Colo et al., 2004b). Such a mucoadhesion could prolong the drug residence at the absorption site. Despite the different effects of Ch-HCl and CMCh on ofloxacin ocular pharmacokinetics, these polymers increased the drug intraocular bioavailability with respect to the reference by about the same factor, as shown by the relevant AUC values. CMCh, being a polyanion, is potentially able to bind cationic drugs. Then it can prolong the precorneal residence of, e.g., aminoglucoside antibiotics at effective antimicrobial concentrations, thus allowing reduction of the frequency of instillations.

Polycationic derivatives of Ch, soluble in tear fluid at the physiological pH of 7.4, can have an increased potential for enhancing the corneal permeability (Di Colo et al., 2004a). A similar derivative, i.e., N-trimethylchitosan (TMC), has been obtained by quaternizing the primary amino group of Ch with methyl iodide, thus bestowing fixed, pH-independent positive charges on the polymer (Fig.4). The complete quaternization is unnecessary to make the polymer soluble at neutral or alkaline pH, in fact, a quaternization degree (QD) of around 40% is sufficient for this purpose (Snyman et al., 2002; Di Colo et al., 2004a). TMC has proved a potent permeation enhancer of hydrophilic molecules and macromolecules across the intestinal epithelium in neutral environments (Kotzé et al., 1997; Kotzé et al., 1998; Thanou et al., 2000). A study of the enhancing effects of TMC polymers having varying QD on the permeation of the hydrophilic mannitol or PEG 4000 across epithelial cell monolayers (Caco-2) demonstrated the existence of an optimum QD, beyond which the permeability was no further increased (Hamman et al., 2003).

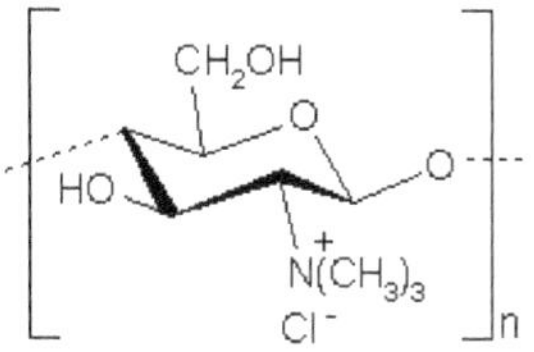

Fig. 4. Structure of N- trimethylchitosan.

It has been shown that TMC is able to enhance the permeability of the rabbit corneal epithelium reconstituted *in vitro* (RRCE) to ofloxacin. A dependence of such a permeabilizing ability on the QD of TMC, similar to that found previously with the Caco-2 intestinal epithelium model (Hamman et al., 2003), has been evidenced (Di Colo et al., 2004a). The steady-state flux of ofloxacin across the RRCE in the presence of TMCs having different QDs was measured and P_{app} values were calculated by dividing such a flux by the drug concentration in the applied solution (0.001% w/v, not cytotoxic to the RRCE cells). The P_{app} enhancement was calculated as the ratio of the P_{app} value in the presence to that in the absence of polymer (enhancement ratio, ER). TMC polymers having low QD (3-4%) were barely soluble and ineffective, whereas those having intermediate QD (35-45%) were soluble and significantly bioactive (ER 1.5-1.6). TMC polymers having high QD (80-90%) showed no further increase of permeability (ER 1.5) (Di Colo et al., 2004a). To explain these findings the following hypotheses may be put forward (Di Colo et al., 2004a; Hamman et al., 2003). At intermediate QD TMC has favorable chain flexibility and conformation whereas at higher QD the TMC electrostatic interactions with the epithelial cell membrane may be hindered by steric effects of the attached methyl groups. Another explanation may be the saturation of the interaction sites on the membrane. A polycationic polysaccharide, i.e., the fully quaternized N-methyl-diethylaminoethyl dextran (MeDEAED) was tested for its ability to permeabilize the RRCE. Although MeDEAED has similar charge density and molecular weight as the TMC having QD 80%, yet only the latter produced a significant P_{app} increase on the RRCE (Di Colo et al., 2004a). This means that charge density and MW are not the sole properties of Chs that concur to determine the ability of these polysaccharides to promote transcorneal drug absorption.

The TMC polymers with intermediate QD, which had proven effective on the RRCE, were also tested *in vivo*, in the eyes of albino rabbits. The polymers were synthesized from Chs of MW 580 kDa (TMCL) and 1460 kDa (TMCH), respectively, in order to investigate the relevance of MW to the polymer bioactivity. The pharmacokinetic data for ofloxacin obtained in the presence of these derivatives were compared with those obtained with Ch-HCl (Di Colo et al., 2004a; Di Colo et al., 2004b). For making these comparisons indicative of the relative bioactivity of polymers all solutions were made isoviscous using PVA, which had been shown to be inert on the cornea (Di Colo et al., 2004b). A significant enhancement of transcorneal absorption rate was produced by TMCH, i.e., the Ch derivative with higher MW, through an increase of corneal permeability, as could be deduced from increases of drug bioavailability and peak concentration in the aqueous (AUC and C_{max}, respectively) by 283 % and 318 %, respectively, over the control. Such an enhancement was also indicated by a shortening of the time to peak (t_{max}). In the presence of TMCH the C_{max} exceeded 4 μg/ml, which is the $MIC_{90\%}$ for the more resistant ocular pathogens (Taravella et al., 1999). On the basis of this consideration, TMCH can be regarded as a potential absorption enhancer to be formulated into ophthalmic ofloxacin solutions for the topical treatment of endophthalmitis. A comparison of the data for TMCH with those for Ch-HCl showed that the permeabilizing effect of the partially quaternized derivative is stronger that that of the hydrochloride of unmodified Ch. This was indeed predicted in the foregoing discussion on the basis of considerations about polymer solubility. Data for TMCL showed that for this derivative C_{max} was significantly lower and t_{max} longer than the corresponding values for TMCH, which points to a weaker enhancing effect of TMCL. The stronger effect of the derivative having the higher MW was ascribed to its stronger adhesion to the corneal mucus (Di Colo et al., 2004a). Indeed it was reported that mucoadhesion, along with the opening of tight junctions, is a key element of TMC polymers for being effective as absorption enhancers at mucosal surfaces (Snyman et al., 2003).

The absorption-enhancing efficacy of TMC was thought to depend on its charge density which, in neutral or alkaline environments, is determined by its quaternization degree (Hamman et al., 2003). In the light of this consideration, novel chitosan derivatives with pendant quaternary ammonium groups were prepared by reacting Ch with 2-diethylaminoethyl chloride (DEAE-Cl) under different conditions (Zambito et al., 2006b). The general structure of these derivatives, assessed by NMR analysis, is depicted in Fig.5.

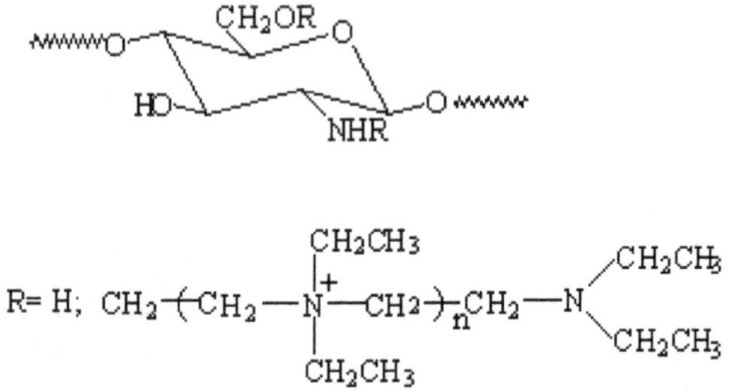

Fig. 5. Structure of N,O-[N,N-diethylaminomethyl(diethyldimethylene ammonium)$_n$methyl] chitosans.

They are assigned the general code, N⁺-Ch, referring to their nature of quaternary ammonium (N^+)-chitosan (Ch) conjugates. The degree of substitution by the pendant chain (DS) and the mean number of quaternary ammonium groups in the chain (n) depend on the MW of the starting Ch and on the molar ratio between reactant (DEAE-Cl) and Ch repeating unit, used in the synthesis. A Ch from shrimp shell (MW=590 kDa), with a reagent excess of 4:1 yielded a derivative, N⁺-Ch-4, with values of the parameters, DS and n, higher than those of N⁺-Ch-2, obtained with a reagent excess of 2:1 (DS=132, n= 2.5 $vs.$ DS=40, n=1.6). The transcorneal penetration-enhancing properties of N⁺-Ch-2 and N⁺-Ch-4 were tested $ex\ vivo$ on the excised rabbit cornea, using the lipophilic dexamethasone and the hydrophilic fluorescein sodium as permeabilization probes (Zambito et al., 2007). The results were compared with corresponding data obtained with a TMC having a QD of 46%, synthesized from the same Ch as that used to prepare the N⁺-Ch conjugates. Dexamethasone was applied on the cornea as a suspension, hence the relevant P_{app} values were calculated by dividing the steady-state flux by the drug aqueous solubility. A significant binding of fluorescein to the Ch derivatives was measured by a dynamic dialysis technique, so the concentration of free fluorescein applied to the cornea was used to calculate the relevant P_{app}.

TMC and the N⁺-Ch derivatives of whichever DS and n, each at the concentration of 1% w/v, enhanced the P_{app} of the hydrophobic dexamethasone across the excised cornea to about the same extent. On the other hand the ER value for the hydrophilic probe fluorescein sodium was the lowest with N⁺-Ch-2 (the less substituted), intermediate with TMC, and the highest with N⁺-Ch-4 (the more substituted). The apparent difference between the polymer effects on the corneal permeability of dexamethasone and fluorescein reflects the difference in the corneal structures on which the effects were probably exerted, that is, the membrane of the corneal cells, in the case of dexamethasone, and the tight junctions connecting the corneal cells, in the case of fluorescein (Zambito et al., 2007).

N⁺-Ch-4 has been tested for its ability to promote the intraocular penetration of dexamethasone or fluorescein $in\ vivo$, in rabbit eyes (Zambito et al., 2007). The intraocular availability of dexamethasone was much higher than that of fluorescein in the reference eye drops. This reflected a much better partitioning of the lipophilic dexamethasone in the corneal cell membrane. In fact, fluorescein transcorneal penetration from the reference eye drops was almost insignificant essentially because of a poor partitioning of the permeant in the cornea. N⁺-Ch-4 significantly enhanced intraocular absorption of fluorescein. In fact the results of the $ex\ vivo$ and $in\ vivo$ experiments agree in indicating a much stronger enhancement effect of this polymer on fluorescein than on dexamethasone. Yet the enhanced P_{app}, AUC and C_{max} values for fluorescein, which reflect the enhanced availability of this highly polar drug model in the aqueous humor, remained lower than the respective values for the non-polar dexamethasone. This means that the paracellular route across the cornea, which is the more likely for very polar molecules, such as fluorescein sodium, remains difficult to penetrate despite the permeabilizing action of the Ch derivatives. On the other hand the use of such enhancers as TMC and N⁺-Ch conjugates may turn profitable for improving the effectiveness of existing commercial ophthalmic formulations of such drugs as dexamethasone and ofloxacin.

As in the cases of protonated or quaternized Chs, the activity of which as epithelial permeability enhancers is now well documented (see, e.g., Di Colo et al., 2008), an electrostatic interaction with negatively charged sites in the mucus, on the cell membranes, or in the tight junctions joining epithelial cells is supposed to be at the basis of the bioactivity of the N⁺-Ch conjugates. The more significant representatives of these derivatives are characterized by degrees of substitution of 40-60% (Zambito et al., 2008). The comparatively high fraction of free, unsubstituted primary amino groups still available on the polymer backbone has been used for covalent attachment of thiol-bearing compounds, via formation of 3-mercaptopropionamide moieties. This has led to water-soluble thiolated quaternary ammonium-Ch conjugates (N$^+$-Ch-SH) the epithelial permeability-enhancing potential of which was tested using the Caco-2 cell monolayer and the excised rat jejunum as substrates (Zambito et al., 2009). On the basis of the obtained results the quaternary ammonium groups of these derivatives were ascribed the ability to reversibly open the epithelial tight junctions and also perturb the plasma membrane of the epithelial cells. On their part the thiol groups were believed to keep the polymer adherent to the epithelium by reacting with the thiols of the epithelial mucus to form disulphide bonds thus favouring the permeability-enhancing action of the positive ions. An N$^+$-Ch conjugate with DS=60% and n=1.7 was used to synthesize a multifunctional non-cytotoxic thiomer carrying 4.5% thiol-bearing 3-mercaptopropionamide besides quaternary ammonium groups (Fig.6).

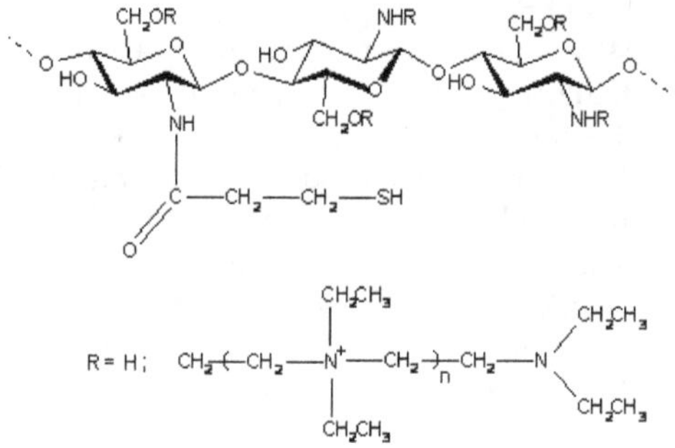

Fig. 6. Strcture of thiolated quaternary ammonium-Ch conjugates

The potential of this N$^+$-Ch-SH thiomer and the parent N$^+$-Ch as bioactive excipients for dexamethasone eye drops was evaluated. The drug permeability across excised rabbit cornea was enhanced over the control value by the thiomer and the parent polymer to about the same extent (3.8 vs. 4.1 times). The mean precorneal retention time (MRT) and AUC in the aqueous of dexamethasone instilled in rabbit eyes *via* eye drops were enhanced by the thiomer (MRT=77.96±3.57 min, AUC=33.19±6.96 µg ml^{-1} min) more than the parent polymer (MRT=65.74±4.91 min, AUC=21.48±3.81 µg ml^{-1} min) over the control (MRT=5.07±0.25 min, AUC=6.25±0.65 µg ml^{-1} min). The quaternary ammonium ions were responsible for both permeabilization of corneal epithelium and polymer adhesion to precorneal mucus, while the thiols increased the latter. This synergistic action is the basis of the higher thiomer bioactivity *in vivo* (Zambito & Di Colo, 2010).

4.1.2 In situ-gelling systems

It is now known that prolonged-release ocular inserts, though causing some discomfort to patients, can remarkably increase the drug ocular bioavailability with respect to the traditional eye drops (Saettone & Salminen, 1995; Di Colo et al., 2001a; Di Colo et al., 2001b). It was found that ocular inserts based on poly(ethylene oxide) (PEO) immediately after application in the lower conjunctival sac of the rabbit eye formed mucoadhesive gels, well tolerated by the animals; then the gels spread over the corneal surface and eroded (Di Colo et al., 2001a; Di Colo et al., 2001b). In order to evaluate the potential of Ch as an intraocular drug absorption promoter ofloxacin was selected as a drug of practical interest and Ch-HCl microspheres medicated with this drug were dispersed into PEO (MW 900 kDa) in the PEO-Ch-HCl (9:1) wt proportion. Erodible ocular inserts, each containing a dose of 0.3 mg ofloxacin, were obtained by compression of the mixture. They were fairly tolerable in the precorneal area of rabbits. The ofloxacin concentration profiles in the aqueous, following administration of a 0.3 mg dose by the PEO-Ch-HCl (9:1) insert, or a medicated PEO insert not containing Ch-HCl, or commercial Exocin eye drops were obtained along with the relevant pharmacokinetic data (C_{max}, t_{max}, AUC) (Di Colo et al., 2002).

As expected, the inserts produced remarkable (one order of magnitude) increases of the AUC values, hence of intraocular bioavailability, over the commercial eye drops. The medicated Ch-HCl microspheres dispersed in the PEO insert produced no substantial bioavailability change with respect to the plain PEO insert. However, with the PEO-Ch-HCl (9:1) insert the C_{max} value was significantly increased over that produced by the Ch-HCl-free PEO insert (7.16±0.77 vs. 4.39±0.58 μg ml^{-1}), while the t_{max} was reduced (150 vs. 300 min). It was reasoned that Ch-HCl, once in contact with the cornea, could exert a gradual enhancing effect on the corneal permeability which could explain the above data (Di Colo et al., 2002).

Microspheres of TMC, medicated with dexamethasone or tobramycin sulfate, were dispersed into PEO (MW 900 kDa) inserts with the aim of both maximizing the intraocular bioavailability of the drugs and studying the permeabilizing effects of TMC on the cornea (Zambito et al., 2006a). Introduction of 10% TMC microspheres into PEO inserts did not substantially affect the drug release pattern and rate from vehicle, nor the vehicle residence time in the precorneal area of rabbits. This allowed assessing the TMC effect on corneal permeability simply by comparing the concentration vs. time profiles in the aqueous obtained with the TMC-containing and the TMC-free inserts. The presence of TMC in the insert significantly increased C_{max} (5.69±0.49 vs. 3.07±0.31 μg ml^{-1}) and AUC (619.3±32.5 vs. 380.5±32.0 μg ml^{-1}min) for a dose of 0.3 mg of the lipophilic dexamethasone, indicating an enhancement effect of TMC on the transcellular penetration pathway. This effect could be exerted through an interaction with the glycoproteins of the mucous layer covering the cornea and/or with the lipid bilayer of the corneal cell membrane (Zambito et al., 2006a). In fact, interactions of protonated chitosan with model phospholipid membranes have been reported (Chan et al., 2001; Fang et al., 2001). A comparison of pharmacokinetic data for an equal dose of the dexamethasone administered as eye drops (C_{max}=0.21±0.02μg ml^{-1}; AUC=17.1±0.1 μg ml^{-1}min) with those obtained with the inserts shows a tremendous potential of inserts for maximizing intraocular drug availability. This compensates for the moderate discomfort in situ gel-forming inserts may cause to patients. On the other hand, the tobramycin concentration in the aqueous was below the limit of detection, even with the TMC-containing insert. It is known that TMC is able to enhance the paracellular penetration of hydrophilic molecules or macromolecules across cell monolayers, such as the intestinal epithelium, by opening the tight junctions between cells (Kotzé et al., 1997; Kotzé et al., 1998; Thanou et al., 2000). The results obtained with tobramycin (Zambito et al., 2006a) demonstrate that the tight junctions of such a stratified epithelium as the cornea are not effectively opened by TMC, at least to such an extent as to allow therapeutically effective penetration of hydrophilic drugs of the tobramycin molecular size. If the cornea is virtually impermeable to tobramycin, although the external cell layer might be affected by TMC the deeper layers could remain impervious to this drug. On the other hand, dexamethasone could effectively permeate the cornea, hence the TMC action, although exerted on the corneal surface, could significantly enhance the apparent permeability of this drug.

In the foregoing discussion it was shown that the transcorneal penetration of a paracellular marker, such as fluorescein sodium, was significantly promoted by the polycationic Ch derivatives TMC and N+-Ch, although the enhanced flux remained comparatively low. These results can be reconciled with those currently discussed concerning tobramycin by considering that the enhancing effect could be measured in the case of fluorescein, but not in that of tobramycin, because the detection limit of the analytical method for the former was much lower than for the latter ($8x10^{-3}$ vs. 0.5 μg/ml). In the light of the discussion so far, the remarkable promotion of the transcorneal absorption of ofloxacin by TMC (Di Colo et al., 2004a) is ascribable to a polymer action on the mucous layer covering the cornea and/or on the corneal cell membrane, rather than to an effective opening of the tight junctions in all layers of corneal cells.

To overcome the bioavailability problems associated with the administration of conventional eye drops the use of *in situ* gel-forming systems that are instilled as drops into the eye and undergo a sol-gel transition in the cul-de-sac has been proposed (Gupta et al., 2007). Different combinations of Ch (MW 150 kDa, 75-85% deacetylated) and poloxamer (Pluronic F-127) were evaluated for gelling ability and viscosity. The selected formulation (0.25% chitosan, 9.0% Pluronic F-127) had a viscosity that could allow easy instillation into the eye as drops and rapid sol-to-gel transition in the precorneal area triggered by a rise in pH from 6.0 to 7.4 and in temperature from ambient to 35-37 °C. A 0.25% timolol concentration usually prescribed for the therapy of glaucoma was added to the formulation. This was a clear, isotonic solution having pH 6.0-6.2. The formulation was converted into a stiff gel during autoclaving but recovered its original properties after cooling. Transcorneal permeation was studied *ex vivo* using excised goat corneas. The authors report a larger drug amount permeated after 4 h with the *in situ* gelling system (63.41±2.6%) as compared with the plain drug solution (42.11±2.1%) (Gupta et al., 2007) . They ascribe this difference to a permeabilizing action of Ch on the cornea, on the basis of the renown of Ch as a transmucosal permeation enhancer. It should be considered, however that the comparison between gelling system and control is based on the cumulative permeation and not on the P_{app} value. Then factors other than corneal permeability might contribute to the difference. Ocular irritation was tested by the chick embryo chorioallantoic membrane test (Gupta et al., 2007). It was found that the *in situ* gel formulation is nonirritant to mildly irritant and is well tolerated. *In vivo* precorneal drainage of the formulation was assessed by gamma scintigraphy, using albino rabbits for the study. The observation of the acquired gamma camera images showed good spreading of the *in situ*-gelling system over the entire precorneal area immediately after administration. The curve of the remaining activity on the corneal surface as a function of time showed that the plain drug solution cleared very rapidly from the corneal region. The drug passed into the systemic circulation, as significant activity was recorded in kidney and bladder 2 h after ocular administration. On the other hand the *in situ* gel formulation cleared slowly and no radioactivity was observed in kidney and bladder. This behaviour was ascribed to the known bioadhesivity of Ch and to the gelation that was induced by Ch and Pluronic F-127 (Gupta et al., 2007).

Another *in situ* gel-forming solution was obtained by coupling poly(N-isopropylacrylamide) (PNIPAAm), a well-known thermosensitive polymer having a thermoreversible phase transition temperature close to human body surface (Maeda et al., 2006; Taylor & Ceranaowski, 1975; Hsiue et al., 2002; Hsiue et al., 2003), with Ch with the hope that this novel polymer (PNIPAAm-Ch) might couple the advantages of Ch and PNIPAAm (Cao et al., 2007). The ocular pharmacokinetics of timolol maleate, applied by this thermosensitive gel-forming system, were measured by microdialysis, a technique for continuously monitoring the drug in the rabbit aqueous (Wei et al., 2006; Macha & Mitra, 2001; Rittenhouse et al., 2000). By means of the sampling technique the behaviour of the drug was studied after the gel-forming solution and the conventional eye drops, both at the concentration of 0.5%, were administered in the *cul-de-sac*. The peak concentrations of 5.58 and 11.2 ng/ml were reached in the aqueous 15 and 30 min after instillation of the conventional eye drops and the thermosensitive gel-forming solution, respectively. It was also noted that the AUC for the latter was two-fold greater than that for the former. These data suggested that the gel-forming system improved the intraocular bioavailability and effectiveness of timolol maleate, which was embedded in the gel and therefore retained in the pre-corneal area for a prolonged period. As a hypothesis from the authors, however not substantiated by further data, PNIPAAm-Ch might enhance the corneal permeability and absorption of timolol maleate due to its positive charge and adhesive characteristics. The MTT assay showed little cytotoxicity of PNIPAAm-Ch at concentrations in the 0.5-400 µg/ml range (Cao et al., 2007).

4.1.3 Ch-based colloidal drug carriers

The potential of colloidal drug carriers, such as liposomes, submicron emulsions, nanocapsules and nanoparticles in ocular delivery has been put into evidence by a review article in 2009 (Alonso et al., 2009). It has been shown that the action of Ch-based nanoparticle systems relies on particle interactivity with the corneal or conjunctival epithelium cells. The nanoparticles can be uptaken within the epithelial cells without causing any damage to them. In this way these nanosystems, once loaded with drugs, might make the conjunctival and corneal epithelia reservoirs for drug delivery to the exterior or interior of the eye. The interaction between Ch nanoparticles and the corneal and conjunctival epithelial surfaces have been investigated (Lehr et al, 1992). The nanoparticles were usually obtained by adding a solution of tripolyphosphate to a Ch solution in acetic acid. To study the interaction of Ch nanoparticles with ocular structures by spectrofluorimetry and confocal fluorescence microscopy, Ch was converted into a fluorescent derivative by the reaction between the fluorescein acid group and the Ch amino group. The stability of nanoparticles in the presence of proteins and enzymes was deemed a crucial issue. The presence of lysozyme was found not to significantly compromise the integrity of Ch nanoparticles since only a slight particle size reduction and no surface charge modification were observed (Lehr et al, 1992). Fluorescence microscopy of eyeball and lid sections confirmed the *in vivo* uptake of Ch nanoparticles by conjunctival and corneal epithelia. *In vivo* studies in rabbits showed that the nanoparticles were well tolerated by the ocular surface structures, which showed no histologic alterations nor abnormal inflammatory cells in cornea, conjunctiva and lids, in full consistency with the lack of clinical signs (de Salamanca et al., 2006).

Another potential colloidal drug carrier has been studied, which combines liposomes and Ch nanoparticles (Diebold et al. , 2007). The rehydration at 60 °C of a lyophilized mixture of Ch nanoparticles, loaded with FITC-BSA, and liposomes led to the coating of nanoparticles with a phospholipid shell. The resulting nanosystem was characterized for size and zeta potential (407.8±9.6 nm and +5.8±1.3 mV to 755.3±30.0 nm and +14.7±0.4 mV, depending on composition) while its structure was assumed theoretically on the basis of an interaction between the positively charged nanoparticles and the negatively charged lipid vesicles, and a reorganization of the membranes to cover the nanoparticle surface. All nanosystems, namely, Ch nanoparticles, liposomes, and liposome-Ch nanoparticle complexes showed physical stability so far as no significant change in particle size was observed by photon-correlation spectroscopy after 2 h in simulated lachrymal fluid. The underlying hypothesis of the authors was that an appropriate combination of liposomes and Ch nanoparticles could increase the ability of the resulting system to interact with biological surfaces and cell membranes and potentially deliver drugs to target tissues.

Studies of Ch-coated colloidal systems loaded with tetanus toxoid, indomethacin or diazepam, or Ch-based nanoparticles loaded with cyclosporin have been reviewed (Alonso & Sanchez, 2009). It was concluded that a Ch coating could add a clear benefit to the potential of colloidal systems as ocular drug carriers, and that Ch nanoparticles might represent an interesting vehicle for drugs the target of which is the ocular mucosa.

The preparation of Ch/Cabopol nanoparticles loaded with pilocarpine, a drug used for the treatment of glaucoma, has been reported (Huei-Jen et al., 2006). A solution of Ch in 1% w/v acetic acid was dropped into a Carbopol dispersion under stirring by a homogenizer to form an opalescent suspension of Ch/Carbopol nanoparticles. The pilocarpine-loaded nanoparticles were prepared by dissolving the drug in a measured volume of Ch/Carbopol nanoparticle suspension and stirring for 48 h. The drug-loaded nanoparticles were isolated by ultracentrifugation and dried by lyophilization. The nanoparticles were assumed to be formed by ionic interaction between the polycationic Ch and the polyanionic Carbopol. A particle size of 294±30 nm, a zeta potential of +55.78±3.41 mV and a pilocarpine encapsulation efficiency of 77±4% were reported (Huei-Jen et al., 2006). After pilocarpine in various formulations, i.e., eye drops, liposomes, gel and nanoparticles, had been applied in the eyes of rabbits, the resulting miotic responses were compared. The decrease in pupil diameter was in the rank order of nanoparticles>liposomes>gel>eyedrops. With nanoparticles and liposomes the miosis effect lasted up to 24 h. The AUC for the curve of decrease in pupil diameter *vs.* time was the largest for the nanoparticle formulation, indicating that this formulation was the most efficient system of the four tested for the topical delivery of pilocarpine (Huei-Jen et al., 2006).

Lipophilic nanoparticles for the delivery of the macrolide rapamycin for immunosuppression in corneal transplantation have recently been described (Xu-Bo et al., 2008). In brief, the preparation of rapamycin-loaded nanoparticles is described by the authors as follows. Poly(lactic acid) (PLA) and rapamycin, dissolved in acetone, are added under ultrasonication to an aqueous solution containing a Ch-cholesterol conjugate (Ch-Chol), after which the solvent is removed by evaporation under stirring. The resulting nanoparticle suspension is centrifuged to remove the drug not entrapped within the particles, and then lyophilized. It is claimed that the amphiphilic Ch-Chol self-aggregates into nanoparticles, with hydrophobic microenvironment inside. When PLA was added to the aqueous Ch-Chol under ultrasonication the size of the nanoparticles slightly increased to about 300 nm, with a zeta potential of +30.3 mV. The presence of PLA favoured the entrapment of the hydrophobic drug into the particles. The drug-loaded Ch-Chol/PLA nanoparticles and a rapamycin suspension were radiolabeled and the ocular distribution of either system was assessed by scintillation counter and single photon emission computed tomography (SPECT) image analysis. The rabbits treated with Ch-Chol/PLA nanoparticles showed radioactivity fractions remaining on cornea and conjunctiva significantly higher than those treated with the suspension of rapamycin. This behaviour is ascribed to the mucoadhesive character of the Ch nanoparticles (Xu-Bo et al., 2008), while disregarding the possibility of particle internalization by corneal and/or conjunctival cells. The radioactivity levels in the aqueous and iris/ciliary body were close to background level. This is taken as an indication that the corneal barrier hindered transport of either the drug or the nanoparticles. Nevertheless the prolonged residence of nanoparticles on the ocular surface is expected to promote drug absorption by the external ocular tissues. The rapamycin-loaded Ch-Chol/PLA were used to treat corneal allografts in comparison with the drug-free nanoparticles and the rapamycin suspension eyedrops. All of 10 grafts in the untreated control group were rejected within 13 days (median survival time, 10.6±1.26 days). Rabbits treated with empty nanoparticles rejected the corneal allografts in a median time of 10.9±1.45 days and none of these grafts survived beyond 13 days. In the group treated with the rapamycin suspension, grafts were rejected between 19 and 27 days with a median survival time of 23.7±3.20 days, while in the group treated with the rapamycin-loaded nanoparticles the median survival time of grafts was 27.2±1.03 days and 50% grafts were still surviving by the end of the observation (4 weeks). These results indicated an improved immunosuppressive effect compared with rapamycin eye drops (Xu-Bo et al., 2008).

A very recent report describes the potential use in ocular drug delivery of liposomes coated with low molecular weight (8 kDa) Ch (LCh) (Li et al., 2009). LCh was prepared by degradation of Ch with H_2O_2. Liposomes loaded with diclofenac sodium were coated with LCh (LChL). These systems, containing 0.1% diclofenac sodium, were compared with a 0.1% aqueous solution of the drug (control) for their effects on drug retention in precorneal area of rabbits. The LChL formulations produced significantly higher AUC (area under concentration in tear fluid vs. time curve) and longer MRT than either non-coated liposomes or the control solution, indicating that the LCh coating was essential to prolong the retention of liposome-encapsulated drug. No irritation or toxicity, caused by continual administration of LChL in a period of 7 days, resulted from an ocular tolerance study. The effect of LChL on drug corneal penetration was studied (Li et al., 2009) using excised rabbit cornea and a diffusion apparatus (Camber, 1985). The apparent corneal permeability was determined by normalizing the permeant steady-state flux to the permeant concentration in the donor (1.0 mg/ml). The P_{app} produced by LChL was significantly higher than that relative to non-coated liposomes, while the latter was not higher than that produced by the aqueous solution. According to the authors (Li et al., 2009) the LCh coating could intensify liposome binding to the corneal surface, thus facilitating drug absorption into the cornea. In addition, the polycationic LCh could enhance the corneal permeability as in the case of the Ch of much higher molecular weight, described earlier (Di Colo et al., 2004b). These results are interesting in that they show that a Ch of as low a molecular weight as 8 kDa can act as a corneal permeabilizer. However, the effect of this Ch on precorneal retention and corneal drug permeability is not reported for the case where drug and Ch are applied in solution, so the advantages of using a liposome formulation instead of a solution are not neatly highlighted by data.

4.2 Xyloglucan

Xyloglucan is a polysaccharide derived from tamarind seeds, therefore it is often coded TSP (tamarind seed polysaccharide). It has a backbone of β(1→4)-linked glucose residues. Three out of four glucose units are substituted with $\alpha(1\rightarrow 6)$ xylose residues, which are partially substituted by $\beta(1\rightarrow 2)$-linked galactose, as shown in Fig.7. Some of the galactose residues can be further substituted with $\alpha(1\rightarrow 2)$ fucose. TSP is highly water-soluble.

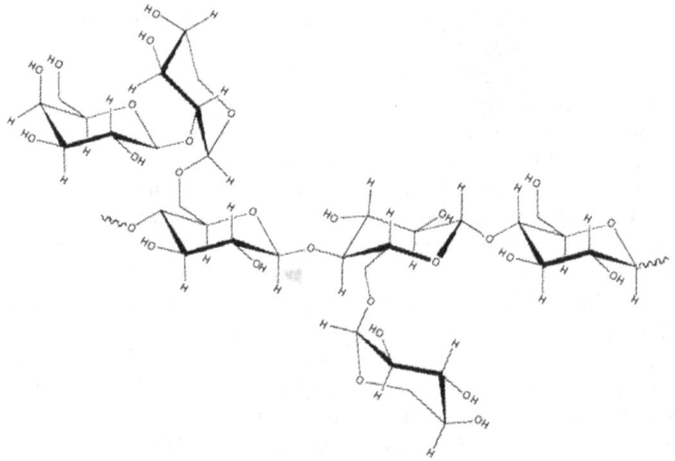

Fig. 7. Structure of the polysaccharide derived from tamarind seeds.

4.2.1 Eye drops

TSP has been described as a viscosity enhancer with mucomimetic activity. Therefore it is currently used in commercial artificial tears for the treatment of dry eye syndrome (DES) (Saettone et al., 1997). Concentrations of such antibiotics as gentamicin and ofloxacin in the rabbit aqueous humor depended on whether rabbit eyes were topically treated with antibiotics alone or drug formulations viscosified with TSP. In the latter instance significantly higher intraocular drug levels were observed. Also, the drugs delivered with TSP produced significantly higher intra-corneal levels than those attained with the corresponding TSP-free formulations. This suggested that TSP enhances corneal drug accumulation by reducing the wash-out of drugs (Ghelardi et al., 2000).

Eye drops containing 3 mg/ml rufloxacin and 10 mg/ml TSP, along with other excipients were topically applied to rabbits for the treatment of experimental *Pseudomonas aeruginosa* and *Staphylococcus aureus* keratitis. Rufloxacin delivered by the polysaccharide reduced *P. aeruginosa* and *S. aureus* in the cornea at a higher rate than that obtained with rufloxacin alone. These results suggested that TSP is able to prolong the precorneal residence time of the antibiotic and enhance drug accumulation in the cornea, thereby increasing the intra-aqueous antibiotic penetration. (Ghelardi et al., 2004).

4.2.2 In situ-gelling systems

When xyloglucan is partially degraded by β-galactosidase the resultant product exhibits thermally reversible gelation in dilute aqueous solutions, which does not occur with native xyloglucan. Gelation is only possible when the galactose removal exceeds 35%. The sol-gel transition temperature was shown to decrease from 40 to 5 °C when the galactose removal ratio increased from 35 to 58% (Nanjawade et al., 2007). Xyloglucan formulations have been studied for ocular delivery of pilocarpine, using Poloxamer 407 as a positive thermosensitive control. The 1.5 wt.% xyloglucan formulation enhanced the miotic response to a degree similar to a 25 wt.% Poloxamer 407 gel (Nanjawade et al., 2007).

4.3 Arabinogalactan

Arabinogalactan (AG), a natural polysaccharide contained in *Larix Occidentalis* (Western Larch), has been found to be biocompatible in the eye, mucomimetic and mucoadhesive (Burgalassi et al., 2007). It is a non-ionic highly branched polysaccharide of the 3,6-β-D-galactan type, the side chains of which consist of β-galactose and β arabinose residues (Gregory & Kelly, 1999).

4.3.1 Eye drops

AG dispersions showed a Newtonian non-viscous behaviour (η=1.6 mPas at 10% w/w concentration) along with good mucoadhesive properties, useful for retention on the eye surface. In fact, a prolonged time of residence in rabbit eyes was ascertained using fluorescein-labeled AG. Five percent w/w AG exerted a good protective effect against the appearance of corneal dry spots. It also reduced significantly the healing time of an experimental corneal lesion. These findings suggested that AG is potentially effective for dry eye protection and in the treatment of corneal wounds (Burgalassi et al., 2007).

4.4 Cellulose derivatives

Cellulose is a polysaccharide consisting of a linear chain of several hundred to over ten thousand β(1→4) linked D-glucose units. The chemical structures of cellulose derivatives used in topical ocular formulations (methyl cellulose, hydroxyethyl cellulose, hydroxypropylmethyl cellulose, sodium carboxymethyl cellulose) are shown in Fig.8.

Fig. 8. Structure of cellulose and its derivatives.

4.4.1 Eye drops

Eye drops containing cellulose derivatives, such as methyl cellulose (MC), hydroxypropylmethyl cellulose (HPMC) or sodium carboxymethyl cellulose (NaCMC), have largely been used as eye lubricants for the treatment of DES. This is characterized by a set of alterations of the eye surface which could relate to tear quality, normal makeup of tear film and alterations in blinking or regular closing of eyelids which entail a reduction of the stability of the tear film and the alteration of the eye surface. HPMC solutions were patented as a semisynthetic substitute for tear-film (Hahnenberger, 1997). When applied, an HPMC solution acts to absorb water, thereby expanding the thickness of the tear film and resulting in extended time of lubricant presence on the cornea and decreased eye irritation (Koroloff et al., 2004). Aside from its widespread commercial and retail availability in a variety of products, HPMC has been used as a 2% solution during surgery to aid in corneal protection and during orbital surgery. Treatment with an isotonic 0.5% solution of NaCMC found in the market (Cellufresh®, Allergan SA, Madrid) produced a significant decrease in the frequency of subjective DES symptoms and improvement of tear film interface stability and corneal surface wettability compared to controls (Bruix et al., 2006).

Hydroxyethyl cellulose (HEC) is used as a viscosity-enhancing agent in ophthalmic formulations to prolong corneal contact time and increase intraocular drug levels.

4.4.2 In situ-gelling systems

Aqueous solutions of MC or HPMC at concentrations in the 1-10 wt.% range are liquid at low temperatures, but form gels upon heating. The transition temperature is between 40 and 50 °C for MC and between 75 and 90 °C for HPMC. Sodium chloride lowers the sol-gel transition temperature of MC to 32-34 °C, while the transition temperature of HPMC can be lowered to about 40 °C by reducing the hydroxypropyl molar substitution (Nanjawade et al., 2007).

Gelation of MC or HPMC solutions is produced by the hydrophobic interaction between methoxy-substituted residues. At low temperatures the macromolecules are hydrated and interact by simple entanglement. At higher temperatures the hydration of polymer chains tends to decrease, chain-chain associations take place, and the system approaches a network structure, corresponding to a sharp rise in viscosity. This sol-gel transition has been exploited to design *in situ* gelling systems having low viscosity at 23 °C and forming soft gels at 37 °C (Nanjawade et al., 2007).

4.5 Hyaluronic acid

Hyaluronic acid (HA) is a high molecular weight, natural and linear polysaccharide composed of β-1,3-N-acetyl glucosamine and β-1,4-glucuronic acid repeating disaccharide units (Fig.9). This biocompatible, nonimmunogenic and biodegradable polymer is one of the most hygroscopic molecules in nature, in fact, HA can hydrate up to 1000 times its dry weight. This property is responsible for enhanced hydration of the corneal surface. Moreover, ocular topical application of formulations based on hyaluronic acid reduces the elimination rate of tear fluid and stabilizes the tear film. This is useful for the treatment of DES (Guillaumie et al., 2010).

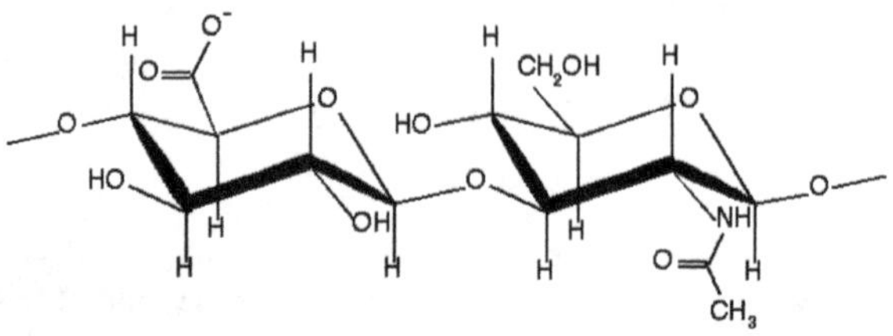

Fig. 9. Structure of hyaluronic acid.

4.5.1 Eye drops

Non-Newtonian and shear thinning properties grant HA solutions a high viscosity at low shear rate (when the eye is open) and a low viscosity at high shear rate (during blinking) thus allowing an even distribution of the solution, improving lubrication of the ocular surface, retarding drainage, improving bioavailability, and reducing discomfort. Another important feature of this high molecular weight, anionic biopolymer is its mucoadhesivity, which provides effective coating and long-lasting protection of the cornea as well as extended residence times on the ocular surface (Guillaumie et al., 2010). Finally, when topically instilled on the eye, hyaluronic acid has been shown to promote physiological wound healing by stimulating corneal epithelial migration and proliferation of keratocytes and to reduce the healing time of corneal epithelium. Topical ophthalmic solutions should exhibit a certain degree of viscosity to prevent immediate drainage from the ocular surface and provide there high efficacy and long residence time. However, the solutions should not be too viscous to avoid blurred vision and to ease their manufacturing process including their sterile filtration (Guillaumie et al., 2010).

4.6 Alginic acid/sodium alginate

Alginic acid (AA) is a natural hydrophilic polysaccharide distributed widely in the cell walls of brown algae. It contains the monomers β-D-mannuronic acid (M) and α-L-guluronic acid (G), arranged as homopolymeric blocks of M-M blocks or G-G blocks together with blocks of alternating sequence (M-G). The polymer forms 3-dimensional ionotropic hydrogel matrices, generally by the preferential interaction of calcium ions with the G moieties resulting in the formation of inhomogeneous gels (Grant et al., 1973 via Cohen et al., 1997).

4.6.1 In situ-gelling systems

An aqueous solution of sodium alginate (NaAA) can gel in the eye. Alginates with G contents of more than 65% instantaneously formed gels upon addition to simulated tear fluid, due to ionotropic gelation by Ca^{++} ions. In contrast, alginates having low G contents formed weak gels at a comparatively low rate (Cohen et al., 1997). Pilocarpine was released slowly from alginate gels *in vitro* over a period of 24 h, and the release occurred mostly via diffusion from the gels. In agreement with the *in vitro* results, intraocular pressure (IOP) measurements of rabbit eyes treated with 2% w/v pilocarpine nitrate in solution or in the *in situ* gel-forming formulation containing high G-content NaAA indicated that the latter significantly extended the duration of the IOP-reducing effect compared to the former (10 h vs. 3 h). On the other hand, the effect of low G-content NaAA on the duration and extent of IOP reduction was insignificant (Cohen et al., 1997).

AA has been evaluated as a potential vehicle in ophthalmic *in situ*-gelling solutions for prolonging the IOP-reducing effect of carteolol. *In vitro* studies indicated that carteolol is released slowly from mucoadhesive AA formulations, suggesting an ionic interaction between the basic drug and AA. IOP measurements of rabbit eyes treated with a 1% carteolol formulation with or without AA showed that this polymer significantly extended the duration of the IOP-reducing effect to 8 h. AA produced an ocular bioavailability increase of a carteolol formulation, as indicated by a drug concentration in the target tissue, following a once-daily administration, equivalent to a twice-daily administration of the AA-free carteolol solution (Sechoy et al., 2000).

4.7 Gellan gum

Gellan gum (GG), commercial name Gelrite®, is a water-soluble polysaccharide produced by the bacterium *Pseudomonas elodea*. The repeating unit of the polymer is a tetrasaccharide consisting of two residues of D-glucose and one of each L-rhamnose and D-glucuronic acid, in the following sequence: [D-Glc($\beta1\rightarrow4$)D-GlcA($\beta1\rightarrow4$)D-Glc($\beta1\rightarrow4$)L-Rha($\alpha1\rightarrow3$)]$_n$.

4.7.1 In situ-gelling systems

GG solutions form gels in the presence of mono- or divalent cations, therefore a 0.6% GG solution containing 0.34% timolol maleate underwent a sol-gel transition in the rabbit eye due to the presence of sodium and calcium ions in tear fluid. *In vitro* release of timolol from GG gelled solutions was retarded and controlled by drug diffusion in the gel. Accordingly, *in vivo* the formation of the gel prolonged the precorneal residence time, and therefore, it increased the ocular bioavailability of timolol in the cornea, aqueous humor and iris+ciliary body of rabbits (Rozier et al., 1989). A similar *in situ*-gelling behaviour and *in vitro* release of the antibacterial agent pefloxacin was reported of a 0.6% GG solution. This allowed attainment of the minimum inhibitory concentration of antibacterial in the aqueous of rabbits, and maintained it for 12 h, while with the conventional eye drops the drug concentration in the aqueous after the same time interval dropped to negligible values (Sultana et al., 2006).

4.8 Polysaccharide mixtures

Mixtures of polysaccharides have been investigated considering that possible interpolimer non-covalent interaction might generate excipients for eye drops, having synergistically improved properties over those of the separate polymers. Ascertaining the above possibility and evaluating the composition of the mixture corresponding to the strongest interaction and optimal biopharmaceutical properties have been the fundamental purposes of these studies.

4.8.1 Eye drops

An interaction between TSP and HA in aqueous solution was ascertained. Various TSP/HA mixtures were studied, among which the 3/2 ratio showed the strongest interaction. The properties of this mixture as a potential excipient for eye drops were synergistically improved over those of the separate polymers. Information about the nature of interpolymer interactions and their dependence on TSP/HA ratio was obtained by nuclear magnetic resonance (NMR) spectroscopy. The affinity of the TSP/HA (3/2) mixture for mucin, assessed by NMR, is higher than that of the single polysaccharides. The mucoadhesivity of this mixture, evaluated *in vitro* by NMR or viscometry, and *in vivo* by its mean and maximum residence time in rabbit precorneal area, is stronger than that of the component polysaccharides or the TSP/HA mixtures of different composition. TSP/HA (3/2) is little viscous and well tolerated by rabbit eyes, so it shows a considerable potential for the treatment of DES. This mixture stabilizes the tear film, hence it has been shown to prolong the residence of drugs, such as ketotifen fumarate and diclofenac sodium, in tear fluid. It is unable to permeabilize the cornea, therefore mucoadhesivity is responsible for the TSP/HA (3/2) synergistic enhancement of either extra- or intra-ocular drug bioavailability (Uccello-Barretta et al., 2010).

4.8.2 In situ-gelling systems

The rheological properties of GG, NaAA and GG/NaAA solutions were evaluated. It was found that the optimum concentration of GG solution for *in situ* gel-forming delivery systems was 0.3% w/w and that of NaAA solution was 1.4% w/w. The mixture of 0.2% GG and 0.6% NaAA, when gelled by simulated tear fluid, showed a significant enhancement in gel strength in physiological conditions. Such a gelled mixture released *in vitro* the anticancer drug matrine most slowly. When the solution of this mixture was instilled in rabbit eyes the *in situ* gelling significantly prolonged drug residence in the precorneal area over the time allowed by the GG or NaAA solution alone (Liu et al., 2010).

A further *in situ* gel-forming polysaccharide mixture is composed of Ch and GG. Both polysaccharides form gels in the physiological conditions of the eye, gelation of Ch being activated by the neutral pH of tear fluid, that of GG by the cations of the electrolytes contained in such a fluid. The *in situ*-gelling system composed of 0.25% w/v Ch and 0.50% w/v GG was prepared by mixing solutions of Ch and GG in acetate buffer pH 5.5-6.0 and ultrapure water, respectively. The system was medicated with 0.25% w/v timolol maleate. *In vitro* transcorneal permeation studies on the formulation showed a significant increase of the drug amount permeated across goat cornea in 4 h compared to plain drug solution and the *in situ*-gelling system based on GG alone. This effect was ascribed to the permeation-enhancing ability of Ch. *In vivo* precorneal drainage of the formulation was studied in rabbits by gamma scintigraphy. Radio-labelled timolol maleate applied as plain solution cleared very rapidly from the corneal region, whereas the drug applied by the Ch/GG *in situ*-gel forming formulation was retained on the corneal surface for a remarkably longer time (Gupta et al., 2010).

4.9 Comparison of polysaccharide mucoadhesivity

Different polysaccharides, i.e., AG, TSP, HA, HEC, have been compared for their ability to resist removal from tear fluid. Their mucoadhesivity was compared *in vitro*, by the polymer-induced viscosity increase of a mucin dispersion, and *in vivo*, by the polymer residence time in rabbit tear fluid (Di Colo et al., 2009).

The optimal polymer to be used as an additive in ophthalmic drops should be mucoadhesive without increasing the viscosity of the solution to an excessive extent. A solution of TSP 0.7% w/v was shown to possess more of these properties than the other polymers at comparison, in fact, it exhibited the highest MRT in the rabbit precorneal area. HA and HEC, although mucoadhesive, increased the solution viscosity to an excessive degree, and this, in addition to worsening the patient compliance, could induce an anomalous reflex tearing with consequent acceleration of precorneal clearance. AG, at the concentrations at which TSP, HA and HEC were tested, exhibited no significant mucoadhesive properties. This polysaccharide, nevertheless, did not increase the solution viscosity, so it showed the potential for being used at much higher concentrations, e.g., 5-10% w/w, at which its mucoadhesivity is significant (Burgalassi et al., 2007). In virtue of its mucoadhesivity, TSP 0.7% w/v is supposed to stabilize the tear film. For this reason the residence of two different drugs, i.e., ketotifen fumarate and diclofenac sodium, in the precorneal area of rabbit eyes was significantly prolonged by this excipient (Di Colo et al., 2009).

5. Conclusion

The polysaccharides, either natural or semi-synthetic, described in the present survey are non-irritant to the eye and show an ample array of possible solutions to formulation issues in ophthalmology. The polysaccharides introduced in eye drops are mucoadhesive and mucomimetic at comparatively low viscosities. Hence they have been used as lubricants of the eye surface to treat DES. In virtue of their mucoadhesivity they stabilize the tear film and prolong the residence of ophthalmic drugs on the eye surface thereby increasing their extra- and intra-ocular bioavailability. Ch and some of its polycationic derivatives, when topically applied *via* eye drops, have shown the additional ability to promote intra-ocular drug absorption by reversibly permeabilizing the cornea. It is in virtue of this peculiar ability and of mucoadhesivity that Ch has been used to prepare nanoparticles for ocular topical application, or added to ophthalmic *in situ*-gelling systems, based on PEO, poloxamer, or PNIPAAm, intended to improve the ocular drug bioavailability. When the basic excipients of *in situ*-gelling systems were polysaccharides, such as AA/NaAA or GG, the basic principles of the drug bioavailability increase were the prolonged retention of gel in precorneal area and the slow drug release from gel.

Most of the *in vivo* studies of ocular formulations employing polysaccharides have used rabbits as the animal model. It must be recognized, however, that the precorneal clearance determined in rabbits is not representative of that in humans, mainly due to differences in blinking frequency. Such differences may be reflected in differences in shear-thinning of tear film, mucoadhesion of polymer and ultimately in drainage of drugs. Nevertheless, the inaccuracies due to blinking differences can be considered to be similar for the different preparations tested, then the rabbit model is deemed robust for those studies where it has been used for comparative purposes.

6. Abbreviation

AA	alginic acid
AG	arabinogalactan
AUC	area under curve
Ch	chitosan
Ch-Chol	chitosan-cholesterol conjugate
Ch-HCl	chitosan hydrochloride
CMCh	N-carboxymethyl chitosan
DEAE-Cl	2-diethylaminoethyl chloride
DES	dry eye syndrome
DS	degree of substitution
ER	enhancement ratio
FITC	fluorescein-isothiocyanate
G	guluronic acid unit in AA
GG	gellan gum
HA	hyaluronic acid
HEC	hydroxyethyl cellulose
HPMC	hydroxypropylmethyl cellulose
IOP	intra ocular pressure
LCh	low molecular weight chitosan
LChL	liposomes coated with LCh
M	mannuronic acid unit in AA
MC	methyl cellulose
MeDEAED	N- methyl-diethylaminoethyl dextran

MRT	mean precorneal retention time
n	mean number of quaternary ammonium groups in the pendant chains of N^+-Ch
NaAA	sodium alginate
NaCMC	sodium carboxymethyl cellulose
N^+-Ch	N,O-[N,N-diethylaminomethyl(diethyldimethylene ammonium)$_n$methyl] chitosan
N^+-Ch-SH	thiolated quaternary ammonium-chitosan conjugate
P_{app}	apparent permeability
PEO	poly(ethylene oxide)
PLA	poly(lactic acid)
PNIPAAm	poly(N-isopropylacrylamide)
PVA	poly(vinyl alcohol)
QD	quaternization degree
RRCE	rabbit reconstituted corneal epithelium
SPECT	single photon emission computed tomography
TMC	N-trimethyl chitosan
TMCL	TMC synthesized from a Ch of MW 580 kDa
TMCH	TMC synthesized from a Ch of MW 1460 kDa
TSP	tamarind seed polysaccharide

7. References

Alonso, M. J. & Sánchez, A. (2009). The potential of chitosan in ocular drug delivery. *J. Pharm. Pharmacol.*, Vol. 55, pp. 1451-1463.

Artursson, P.; Lindmark, T.; Davis, S.S. & Illum, L. (1994). Effect of chitosan on the permeability of monolayers of intestinal epithelial cells (Caco-2). *Pharm. Res.*, Vol. 11, pp. 1358-1361.

Bruix, A; Adan, A. & Casaroli-Marano, R.P. (2006). Efficacy of sodium carboxymethylcellulose in the treatment of dry eye syndrome. *Arch. Soc. Esp. Oftalmol.*, Vol. 81, pp. 85-92.

Burgalassi, S.; Nicosia, N.; Monti, D.; Falcone, G.; Boldrini, E. & Chetoni, P. (2007). Larch arabinogalactan for dry eye protection and treatment of corneal lesions: Investigation on rabbits. *J. Ocul Pharmacol. Ther.*, Vol. 23, pp. 541-549.

Camber, O. (1985). An *in vitro* model for determination of drug permeability through the cornea. *Acta Pharm. Suec.*, Vol. 22, pp. 335-342.

Cao, Y.; Zhang, C.; Shen, W.; Cheng, Z.; Yu, L.L. & Ping, Q. (2007). Poly(N-isopropylacrylamide)—chitosan as thermosensitive *in situ* gel-forming system for ocular drug delivery. *J.Control. Rel.*, Vol. 120, pp. 186-194.

Chan, V.; Mao, H.Q. & Leong, K.W. (2001). Chitosan-induced perturbation of dipalmitoyl-*sn*-glycero-3-phosphocholine membrane bilayer. *Langmuir*, Vol. 44, pp. 201-208.

Cohen, S.; Lobel, E.; Trevgoda, A.; Peled, Y. (1997). A novel in situ-forming ophthalmic drug delivery system from alginates undergoing gelation in the eye. *J. Control. Rel.*; Vol. 44, pp. 201-208.

de Salamanca, A.E.; Diebold, Y.; Calonge, M.; García-Vazquez, C.; Callejo S.; Vila, A. & Alonso, M.J. (2006). Chitosan nanoparticles as a potential drug delivery system for the ocular surface: toxicity, uptake mechanism and *in vivo* tolerance. *Invest. Ophthalmol. Vis. Sci.*, Vol. 47, pp. 1416-1425.

Diebold, Y.; Jarrín, M.; Sáez, V.; Carvalho, E.L.S.; Orea, M.; Calonge, M.; Seijo, B. & Alonso, M. J. (2007). Ocular drug delivery by liposome-chitosan nanoparticle complexes (LCS-NP). *Biomaterials*, Vol. 28, pp. 1553-1564.

Di Colo, G.; Burgalassi, S.; Chetoni, P.; Fiaschi, M.P.; Zambito, Y. & Saettone, M.F. (2001a). Gel-forming erodible inserts for ocular controlled delivery of ofloxacin. *Int. J. Pharm.*, Vol. 215, pp. 101-111.

Di Colo, G.; Burgalassi, S.; Chetoni, P.; Fiaschi, M. P.; Zambito, Y. & Saettone, M.F. (2001b). Relevance of polymer molecular weight to the *in vitro/in vivo* performances of ocular inserts based on poly(ethylene oxide). *Int. J. Pharm.*, Vol. 220, pp. 169-177.

Di Colo, G.; Zambito, Y.; Burgalassi, S.; Serafini, A. & Saettone, M.F. (2002). Effect of chitosan on in vitro release and ocular delivery of ofloxacin from erodible inserts based on poly(ethylene oxide). *Int. J. Pharm.*, Vol. 248, pp. 115-122.

Di Colo, G.; Burgalassi, S.; Zambito, Y.; Monti, D. & Chetoni, P. (2004a). Effects of different N-trimethylchitosans on *in vitro/in vivo* ofloxacin transcorneal permeation. *J. Pharm. Sci.*, Vol. 93, pp. 2851-2862.

Di Colo, G.; Zambito, Y.; Burgalassi, S.; Nardini, I. & Saettone, M.F. (2004b). Effect of chitosan and of N-carboximethylchitosan on intraocular penetration of topically applied ofloxacin. *Int. J. Pharm.*, Vol. 273, pp. 37-44.

Di Colo, G.; Zambito, Y. & Zaino, C. (2008). Polymeric enhancers of mucosal epithelia permeability: synthesis, transepithelial penetration-enhancing properties, mechanism of action, safety issues. *J. Pharm. Sci.*, Vol. 97, pp. 1652-1680.

Di Colo, G.; Zambito, Y.; Zaino, C. & Sansò, M. (2009). Selected polysaccharides at comparison for their mucoadhesiveness and effect on precorneal residence of different drugs in the rabbit model. *Drug Dev. Ind. Pharm.*, Vol. 35, pp. 941-949.

Dodane, V. & Vilivalam, V.D. (1998). Pharmaceutical applications of chitosan. *Pharma Sci. Technol. Today*, Vol. 1, pp. 246-253.

Fang, N.; Chan, V.; Mao, H. Q. & Leong, K. W. (2001). Interactions of phospholipid bilayer with chitosan: effect of molecular weight and pH. *Biomacromolecules*, Vol. 2, pp. 1161-1168.

Felt, O.; Buri, P. & Gurny, R. (1998). Chitosan: a unique polysaccharide for drug delivery. *Drug Dev. Ind. Pharm.*, Vol. 24, pp. 979-993.

Ghelardi, E.; Tavanti, A.; Pelandroni, F.; Lupetti, A.; Blandizzi, C.; Boldrini, E.; Campa, M. & Senesi M. (2000). Effect of a novel mucoadhesive polysaccharide obtained from tamarind seeds on the intraocular penetration of gentamicin and ofloxacin in rabbits. *J. Antimicrob. Chemother.*, Vol. 48, pp. 3396-3401.

Ghelardi, E.; Tavanti, A.; Davini, P.; Pelandroni, F.; Solvetti, S.; Parisio, E.; Boldrini, E.; Senesi, S. & Campa, M. (2004). A mucoadhesive polymer extracted from tamarind seed improves the intraocular penetration and efficacy of rufloxacin in topical treatment of experimental bacterial keratitis. *Antimicrob. Agents Chemother.*, Vol. 48, pp. 3396-3401.

Grant, G.T.; Morris, E.R.; Rees, D.A.; Smith, P.J.C. & Thom, D. (1973). Biological interactions between polysaccharides and divalent cations: The egg box model. *FEBS Lett.*, Vol. 32, pp. 195-198.

Gregory, S. & Kelly, N.D. (1999). Larch arabinogalactan: Clinical relevance of a novel immune-enhancing polysaccharide. *Altern. Med. Rev.*, Vol. 4, pp. 96-103.

Guillaumie, F.; Furrer, P.; Felt-Baeyens, O.; Fuhlendorff, B.L.; Nymand, S.; Westh, P.; Gurny, R. & Schwach-Abdellaoui, K. (2010). Comparative studies of various hyaluronic acids produced by microbial fermentation for potential topical ophthalmic applications. *J. Biomed. Mater. Res. Part A*, Vol. 92, No. 4, pp. 1421-1430.

Gupta, H.; Jain, S.; Mathur, R.; Mishra, P. & Mishra, A. K. (2007). Sustained ocular drug delivery from a temperature and pH triggered *in situ* gel system. *Drug Deliv.*, Vol. 14, pp. 507-515.

Gupta, H.; Velpandian, T. & Jain, S. (2010). Ion- and pH-activated novel *in-situ* gel system for sustained ocular drug delivery. *J. Drug Targeting*, Vol. 18, pp. 499-505.

Hahnenberger, R.W. (1997). Pharmaceutical composition containing carbachol and other cholinergic substances. United States Patent 5,679,713.

Hamman, J.H.; Schultz, C. M. & Kotzé, A.F. (2003). N-trimethyl chitosan chloride: optimum degree of quaternization for drug absorption enhancement across epithelial cells. *Drug Dev. Ind. Pharm.*, Vol. 29, pp. 161-172.

Huei-Jen, K.; Hong-Ru, L.; Yu-Li, L. & Shi-Ping Y. (2006). Characterization of pilocarpine-loaded chitosan/Carbopol nanoparticles. *J. Pharm. Pharmacol.*, Vol. 58, pp. 79-186.

Hui H.W. & Robinson, J. R.(1985). Ocular drug delivery of progesterone using a bioadhesive polymer. *Int. J. Pharm.*, Vol. 26,pp. 203-213.

Illum, L. (1998). Chitosan and its use as a pharmaceutical excipient. *Pharm. Res.*, Vol. 15, pp. 1326-1331.

Illum, L.; Farraj, N. F. & Davis, S.S. (1994). Chitosan as a novel nasal delivery system for peptide drugs. *Pharm. Res.*, Vol. 11, pp. 1186-1189.

Janes, K.A.; Calvo, P. & Alonso, M.J. (2001). Polysaccharide colloidal particles as delivery systems for macromolecules. *Adv. Drug Deliv. Rev.*, Vol. 47, pp. 83-97.

Koroloff, N.; Boots, R.; Lipman, J.; Thomas, P.; Rickard, C. & Coyer, F. (2004). A randomised controlled study of the efficacy of hypromellose and Lacri-Lube combination versus polyethylene/Cling wrap to prevent corneal epithelial breakdown in the semiconscious intensive care patien". *Intensive Care Med*, Vol. 30, No. 6, pp. 1122-1126.

Kotzé, A.F.; Lueßen, H.L.; De Leeuw, B.J.; De Boer, A.G.; Verhoef, J.C. & Junginger, H.E. (1997). N-trimethyl chitosan chloride as a potential absorption enhancer across mucosal surfaces: *in vitro* evaluation in intestinal epithelial cells (Caco-2). *Pharm. Res.*, Vol. 14, pp. 1197-1202.

Kotzé, A.F.; Lueßen, H.L.; De Leeuw, B.J.; De Boer, A.G.; Verhoef, J.C. & Junginger, H.E. (1998). Comparison of the effect of different chitosan salts and N-trimethyl chitosan chloride on the permeability of intestinal epithelial cells (Caco-2). *J. Control. Rel.*, Vol. 51, pp. 35-46.

Lee, V.H.L. & Robinson, J.R. (1986). Review: topical ocular drug delivery: recent developments and future challenges. *J. Ocul. Pharmacol.*, Vol. 2, pp. 67-108.

Lehr, C.M.; Bowstra, J.A.; Schacht, E.H. & Junginger, H.E. (1992). *In vitro* evaluation of mucoadhesive properties of chitosan and some other natural polymers. *Int. J. Pharm.*, Vol. 78, pp. 43-48.

Li, N.; Zhuang, C.; Wang, M.; Sun, X.; Nie, S. & Pan, W. (2009). Liposome coated with low molecular weight chitosan and its potential use in ocular drug delivery. *Int. J. Pharm.*, Vol. 379, pp. 131-138.

Liu, Y.; Liu, J.; Zhang, X.; Huang, Y. & Wu, C. (2010). *In situ* gelling gelrite/alginate formulations as vehicles for ophthalmic drug delivery. *AAPS Pharm. Sci. Technol.*, Vol. 11, pp. 610-620.

Ludwig, A. (2005). The use of mucoadhesive polymers in ocular drug delivery. *Adv. Drug Del. Rev.*, Vol. 57, pp. 1595-1639.

Macha, S. & Mitra, A. K. (2001). Ocular pharmacokinetics of cephalosporins using microdialysis. *J. Ocular Pharmacol. Ther.*, Vol. 17, pp. 485-498.

Maeda, T.; Kanda, T.; Yonekura, Y.; Yamamoto, K. & Aoyagi, T. (2006). Hydroxylated poly(N-isopropylacrylamide) as functional thermoresponsive materials. *Biomacromolecules*, Vol. 7, pp. 545-549.

Majumdar, S.; Hippalgaonkar, K. & Repka, M.A. (2008). Effect of chitosan, benzalkonium chloride and ethylenediaminotetraacetic acid on permeation of acyclovir across isolated rabbit cornea. *Int. J. Pharm.*, Vol. 348, pp. 175-178.

Muzzarelli, R.A.A.; Tanfani, F.; Emmanueli, M. & Mariotti, S. (1982). N-(carboxymethylidene)-chitosans and N-(carboxymethyl)-chitosans: novel chelating polyampholytes obtained from chitosan glyoxylate. *Carbohydr. Res.*, Vol. 107, pp. 199-214.

Nagarwal, R. C.; Kant, S.; Singh, P.N.; Maiti, P. & Pandit, J.K. (2009). Polymeric nanoparticulate system: A potential approach for ocular drug delivery. *J. Control. Rel.*, Vol. 136, pp. 2-13.

Nanjawade, B.V.; Manvi, F.V. & Manjappa, A.S. (2007). In situ-forming hydrogels for sustained ophthalmic drug delivery. *J. Control. Rel.*, Vol. 122, pp. 119-134.

Paul, W. & Sharma, C. (2000). Chitosan, a drug carrier for the 21st century. *STP Pharma Sci.*, Vol. 10, pp. 5-22.

Rittenhouse, K.D.; Peiffer, R.L. Jr. & Pollack, G.M. (1999). Microdialysis evaluation of the ocular pharmacokinetics of propanolol in the conscious rabbit. *Pharm. Res.*, Vol. 16, pp. 736-742.

Rittenhouse, K.D. & Pollack, G.M. (2000). Microdialysis and drug delivery to the eye. *Adv. Drug Deliv. Rev.*, Vol. 45, pp. 229-241.

Rozier, A.; Mazuel, C.; Grove, J. & Plazonnet, B. (1989). Gelrite®: A novel, ion-activated, in-situ gelling polymer for ophthalmic vehicles. Effect on bioavailability of timolol. *Int. J. Pharm.*, Vol. 57, pp. 163-168

Saettone, M. F. & Salminen, L. (1995). Ocular Inserts for topical delivery. *Adv. Drug Deliv. Rev.*, Vol. 16, pp. 95-106.

Saettone, M.F.; Burgalassi, S.; Boldrini, E.; Bianchini, P. & Luciani, G. (1997). Ophthalmic solutions viscosified with tamarind seed polysaccharide. International patent application. PCT/IT97/00026.

Sechoy, O.; Tissie, G.; Sebastian, C.; Maurin, F.; Driot, J.Y. & Trinquand, C. (2000). A new long acting ophthalmic formulation of carteolol containing alginic acid. *Int. J. Pharm.*, Vol. 207, pp. 109-116.

Singla, A. K. & Chawla, M. (2001). Chitosan: some pharmaceutical and biological aspects – an update. *J. Pharm. Pharmacol.*, Vol. 53, pp. 1047-1067.

Snyman, D.; Hamman, J.H.; Kotzé, J.S.; Rollings, J.E. & Kotzé, A.F. (2002). The relationship between the absolute molecular weight and the degree of quaternisation of N-trimethyl chitosan chloride. *Carbohydrate Polym.*, Vol. 50, pp. 145-150.

Snyman, D.; Hamman, J.H. & Kotzé, A.F. (2003). Evaluation of the mucoadhesive properties of N-trimethyl chitosan chloride. *Drug Dev. Ind. Pharm.*, Vol. 29, pp. 61-69.

Sultana, Y.; Aquil, M. & Ali, A. (2006). Ion-activated Gelrite®-based in situ ophthalmic gels of pefloxacin mesylate: Comparison with conventional eye drops. *Drug Delivery*, Vol. 13, pp. 215-219.

Taravella, M.J.; Balentine, J.; Young, D.A. & Stepp, P. (1999). Collagen shield delivery of ofloxacin to the human eye. *Cataract Refract. Surg.*, Vol. 25, pp. 562-565.

Taylor, L.D. & Ceranaowski, L.D. (1975). Preparation of film exhibiting a balanced temperature dependence to permeation by aqueous solutions-A study of lower consolute behavior. *J. Polym. Sci., Polym. Chem. Ed.*, Vol. 13, pp. 2551-2570.

Thanou, M.; Florea, B.I.; Langemeyer, M.W.E.; Verhoef, J.C. & Junginger, H.E. (2000). N-trimethyl chitosan chloride (TMC) improves the intestinal permeation of the peptide drug buserelin *in vitro* (Caco-2 cells) and *in vivo* (rats). *Pharm. Res.*, Vol. 17, pp. 27-31.

Thanou, M.; Nihot, M.T.; Jansen, M.; Verhoef, J.C. & Junginger, H.E. (2001). Mono-N-carboxymethyl chitosan (MCC), a polyampholytic chitosan derivative, enhances the intestinal absorption of low molecular weight heparin across intestinal epithelia *in vitro* and *in vivo*. *J. Pharm. Sci.*, Vol. 90, pp. 38-46.

Uccello-Barretta, G.; Nazzi, S.; Zambito, Y.; Di Colo, G.; Balzano, F. & Sansò, M. (2010). Synergistic interaction between TS-polysaccharide and hyaluronic acid: Implications in the formulation of eye drops. *Int. J. Pharm*, Vol. 395, pp. 122-131.

Xu-Bo, Y.; Yan-Bo, Y.; Wei, J.; Jie, L.; En-Jiang, T.; Hui-Ming, S.; Ding-Hai, H.; Xiao-Yan, Y.; Hong, L. S. & Jing., S. (2008). Preparation of rapamycin-loaded chitosan/PLA nanoparticles for immunosuppression in corneal transplantation. *Int. J. Pharm.*, Vol. 349, pp. 241-248.

Wei, G.; Ding, P.T.; Zheng, J.M. & Lu, W.Y. (2006). Pharmacokinetics of timolol maleate in aqueous humor sampled by microdialysis after topical administration of thermosetting gels. *Biomed. Chromatogr.*, Vol. 20, pp. 67-71.

Winfield, A.J.; Jessiman, D.; Williams, A. & Esakowitz, L. (1990). A study of the causes of non-compliance by patients prescribed eyedrops. *Br. J. Ophthalmol.*, Vol. 74, pp. 477-480.

Zambito, Y.; Zaino, C. & Di Colo, G. (2006a). Effects of N-trimethylchitosan on transcellular and paracellular transcorneal drug transport. *Eur. J. Pharm. Biopharm.*, Vol. 64, pp. 16-25.

Zambito, Y.; Uccello-Barretta, G.; Zaino, C.; Balzano, F. & Di Colo, G.(2006b). Novel transmucosal absorption enhancers obtained by aminoalkylation of chitosan. *Eur. J. Pharm. Sci.*, Vol. 29, pp. 460-469.

Zambito, Y.; Zaino, C.; Burchielli, S.; Carelli, V.; Serafini, M. F. & Di Colo, G. (2007). Novel quaternary ammonium chitosan derivatives for the promotion of intraocular drug absorption. *J. Drug Deliv. Sci. Technol.*, Vol. 17, pp. 19-24.

Zambito, Y.; Zaino, C.; Uccello-Barretta, G.; Balzano, F. & Di Colo, G. (2008). Improved synthesis of quaternary ammonium-chitosan conjugates (N+-Ch) for enhanced intestinal drug permeation. *Eur. J. Pharm. Sci.*, Vol. 33, pp. 343-350.

Zambito, Y.; Fogli, S.; Zaino, C.; Stefanelli, F.; Breschi, M.C. & Di Colo, G. (2009). Synthesis, characterization and evaluation of thiolated quaternary ammonium-chitosan conjugates for enhanced intestinal drug permeation. *Eur. J. Pharm. Sci.*, Vol. 38, pp. 112-120.

Zambito, Y. & Di Colo, G. (2010). Thiolated quaternary ammonium-chitosan conjugates for enhanced precorneal retention, transcorneal permeation and intraocular absorption of dexamethasone. *Eur. J. Pharm. Biopharm.*, Vol. 75, pp. 194-199.

4

Natural-Based Polyurethane Biomaterials for Medical Applications

Doina Macocinschi, Daniela Filip and Stelian Vlad
Institute of Macromolecular Chemistry "Petru Poni" Iasi
Romania

1. Introduction

Biomaterials are biologically inert or compatible materials placed inside a patient on a long-term or permanent basis. Advances in engineering and a greater availability of synthetic materials triggered the development of engineered polymers for use in biomaterials medical devices. As for other biomaterials, the basic design criteria for polymers used in the body call for compounds that are biocompatible, processable, sterilizable and capable of controlled stability or degradation in response to biological conditions.

The interdisciplinary field of biomaterials and tissue engineering has been one of the most dynamic disciplines during the last decades (Durairaj, 2001; Grundke, 2005; Norde, 2006; Ohya, 2002; Pilkey, 2005; Thomson, 2005; Vermette et al., 2001). The selection of the materials used in the construction of prostheses and implants is basically focused on their ability to maintain mechanical, chemical and structural integrity and on various characteristics which allow this function to substitute any organ or tissue properly and exhibit safe, effective performance within the body. Biocompatibility has been defined as the ability of a material to perform with an appropriate host response in a specific application. Any material used satisfactorily in orthopedic surgery may be inappropriate for cardiovascular applications because of its thrombogenic properties. Any deleterious effects may be encountered if used under stress-strain conditions. Biocompatibility of a material can be simulated by comparing its behaviour to reference materials in standardized experimental condition. Biocompatibility is in fact, complex being interpreted as a series of events of interactions happening at the tissue/material interface, allowing the identification of those materials with surface characteristics and/or polymer chemistry more biocompatible; these interactions are influenced by intrinsic characteristics of the material, the confrontational circumstances related to the bioactive and biocooperative responses.

Blood contact assays have been developed and include tests investigating the adhesion or activation of blood cells, proteins, and macromolecules such as those found in the complement or coagulation cascade. Other biocompatibility tests have been tentatively proposed and involve analytical testing or observations of physiological phenomena, reactions or surface properties attributable to a specific application such as protein adsorption characteristics.

Polyurethanes are one of the most popular groups of biomaterials applied for medical devices. Their segmented block copolymeric character endows them a wide range of versatility in terms of tailoring their physical properties, blood and tissue compatibility.

Polymers from natural sources are particularly useful as biomaterials and in regenerative medicine, given their similarly to the extracellular matrix and other polymers in the human body. Polyester- and polyether-urethanes have been modified with hydroxypropyl cellulose aiming the change of their surface and bulk characteristics to confer them biomaterial qualities (Macocinschi et al., 2008; Macocinschi et al., 2009a; Macocinschi et al., 2009b). In this respect, dynamic contact angle measurements, dynamic mechanical analyses accompanied by mechanical testing have been done. Platelet adhesion test has been carried out *in vitro* and the use of hydroxypropyl cellulose in the polyurethane matrix reduces the platelet adhesion and therefore recommends them as candidates for biocompatible materials. Polymeric composites prepared by mixing polyurethanes and natural polymers offer improved mechanical properties and biocompatibility for functional tissue replacements *in vivo*. The biological characteristics in contact with blood and tissues for long periods, in particular good antithrombogenic properties, recommend the use of extracellular matrix components such as collagen, elastin and glycosaminoglycans (GAG) for obtaining biomaterials (Macocinschi et al., 2010a; Macocinschi et al., 2011; Moldovan et al., 2008; Musteata et al., 2010; Raschip et al., 2009). The introduction of biodegradable polymers into a synthetic polymer matrix restricts the action of a fungal, microbial or enzymatic attack (Macocinschi et al., 2010b). Such limitations appear even when the biodegradable component occurs as a continuous phase in the composite material.

Our previous publications presented the synthesis and some properties of new polyurethane-cellulose, (Macocinschi et al., 2008).

Herein the effects of the chemical structure of polyurethanes-cellulose on their surface properties are discussed. Investigations are based on the geometric mean approach of Kälble, Owens and Wendt, Rabel (Kälble, 1969; Owens & Wendt, 1969; Rabel, 1977), on the Lifshitz-van der Waals acid/base approach of Van Oss and co-workers (Van Oss et al., 1988a; Van Oss et al., 1988b; Van Oss, 1994) and on the theoretical methods involving quantitative structure-property relationship (Bicerano, 1996). By scanning electron microscopy surface morphology was investigated. For estimation of the haemocompatibility properties of the obtained materials, water sorption was determined as well as the amount of fibrinogen adsorbed from solution, the amount of fibrinogen adsorbed from blood plasma, and the time of prothrombin consumption.

2. Surface, mechanical and biological characterizations of polyurethane biomaterials

An ideal polymer for medical applications would have adequate surface and mechanical properties to match the application, would not induce inflammation or other toxic response, and would be sterilizable and easily processed into a final end product with an acceptable shelf life. Polyurethanes are good biomaterials due to their high biocompatibility having chemical structure similar to that of proteins and elastomer characteristics.

2.1 Surface properties

Segmented polyurethanes have gained considerable position as useful biomaterials for implants or biomedical devices, (Baumgartner et al., 1997; Hung et al., 2009; Reddy et al., 2008; Wu et al., 2009). Polyurethanes have been widely used for various commercial and experimental blood contacting and tissue-contacting application, such as vascular prostheses, blood pumps end tracheal tubes, mammary prostheses, heart valves, pacemaker lead wire insulations, intra-aortic balloons, catheters and artificial hearts, because of their generally favorable surface physical properties, together with their fairly good biocompatibility and haemocompatibility characteristics, (Lamba et al., 1997; Lelah & Cooper, 1986; Plank et al., 1987). The balance between the surface hydrophilic and hydrophobic properties is important for achieving an enhanced biocompatibility of polyurethanes. Plasma treatments or other types of stimuli may alter the surface energy of most polymers, thus changing their surface polarity, hydrophilicity and adhesiveness, (Desai et al., 2000; Ozdemir et al., 2002; Ramis et al., 2001).

In Table 1 are provided the compositional parameters and the average molecular weights values, polydispersity indices (GPC).

Sample code	Composition macrodiol/MDI/EG/HPC, wt %	M_n	M_w/M_n
PEA-PU	55.56/37.50/6.94/0.0	109613	1.287
PEA-HPC	52.24/36.57/7.27 /3.92	134522	1.865
PTHF-HPC	52.24/36.57/7.27 /3.92	70291	1.590
PPG-HPC	52.24/36.57/7.27 /3.92	72951	1.669

Table 1. Compositional parameters, number-average molecular weights, polydispersity indices

The measurement methods used for determination of surface tension are based on contact angles between the liquid meniscus and the polyurethane surface. The contact angle is a measure of the ability of a liquid to spread on a surface. The contact angle is linked to the surface energy and so one can calculate the surface energy and discriminate between polar and apolar interactions. Table 2 lists the contact angles between double distilled water, ethylene glycol, or CH_2I_2 and polyurethane samples, before and after effecting of high frequency cold plasma treatment.

Polymer code	Untreated samples/ Plasma –treated samples		
	Water	Ethylene glycol	CH_2I_2
PEA-HPC	45/60	40/35	40/29
PTHF-HPC	45/52	36/29	33/29
PPG-HPC	60/50	45/30	32/27

Table 2. Contact angle degrees of different liquids and polyurethane samples before and after plasma treatment

For the calculation of the surface tension parameters, the geometric mean method (Eqns. (1) and (2)), (Kälble, 1969; Owens & Wendt, 1969; Rabel, 1977) the acid/base method (LW/AB) (Eqns. (3)-(5)), (Van Oss et al., 1988a; Van Oss et al., 1988b; Van Oss, 1994), and theoretical method based on the structure-property relationship considering the group contribution techniques (Eqn. (6)), (Bicerano, 1996) , were used.

$$\frac{1+\cos\theta}{2}\frac{\gamma_{lv}}{\sqrt{\gamma_{lv}^d}}=\sqrt{\gamma_{sv}^p}\cdot\sqrt{\frac{\gamma_{lv}^p}{\gamma_{lv}^d}}+\sqrt{\gamma_{sv}^d} \tag{1}$$

$$\gamma_{sv}=\gamma_{sv}^d+\gamma_{sv}^p \tag{2}$$

where θ is the contact angle determined for water, ethylene glycol and CH_2I_2, subscripts 'lv' and 'sv' denote the interfacial liquid-vapour and surface-vapour tensions, respectively, while superscripts 'p' and 'd' denote the polar and disperse components, respectively, of total surface tension, γ_{sv}.

$$1 + \cos\theta = \frac{2}{\gamma_{lv}}(\sqrt{\gamma_{sv}^{LW} \cdot \gamma_{lv}^{LW}} + \sqrt{\gamma_{sv}^{+} \cdot \gamma_{lv}^{-}} + \sqrt{\gamma_{sv}^{-} \cdot \gamma_{lv}^{+}} \tag{3}$$

$$\gamma_{sv}^{AB} = 2\sqrt{\gamma_{sv}^{+} \cdot \gamma_{sv}^{-}} \tag{4}$$

$$\gamma_{sv}^{LW/AB} = \gamma_{sv}^{LW} + \gamma_{sv}^{AB} \tag{5}$$

where superscripts 'LW' and 'AB' indicate the disperse and the polar component obtained from the γ_{sv}^{-} electron donor and the γ_{sv}^{+} electron acceptor interactions, while superscript 'LW/AB' indicates the total surface tension.

$$\gamma\,(298\text{ K}) \approx 0.75 \cdot [E_{coh}/V(298\text{ K})]^{2/3} \tag{6}$$

where γ is the total surface tension, E_{coh} the cohesive energy and V the molar volume.
According to the geometric mean method, the solid surface tension components were evaluated with Eqn. (1), (Van Oss et al., 1989) using the known surface tension components, (Erbil, 1997; Rankl et al., 2003; Strom et al., 1987) of different liquids from Table 3 and the contact angles from Table 2. The total surface tension was calculated with Eqn. (2).

Test liquids	γ_{lv}	γ_{lv}^{d}	γ_{lv}^{p}	γ_{lv}^{-}	γ_{lv}^{+}
Water	72.8	21.8	51.0	25.5	25.50
Ethylene glycol	48.0	29.0	19.0	47.0	1.92
Methylene Iodide	50.8	50.8	0.0	0.0	0.72

Table 3. Surface tension parameters (mN/m) of the liquids used for contact angle measurements

Table 4 shows the surface tension parameters for both untreated and plasma-treated polyurethane samples, according to the geometric mean method and to the acid/base method. In this table it was considered that γ_{sv}^{LW} is equivalent to γ_{sv}^{d} of the geometric mean method, the mean values of γ_{sv}^{-} and γ_{sv}^{+} were calculated with Eqn. (3). Also, the total surface tension was calculated with Eqn. (2). Following the plasma treatment the disperse component of surface tension, γ_{sv}^{d}, increases in absolute value, while the polar component surface tension γ_{sv}^{p}, decreases except PPG-HPC sample for which these dependences varies in a less extent (γ_{sv}^{p} increases from 32.2 to 38.9 mN/m, and γ_{sv}^{d} increases from 9.1 to 10.7 mN/m).
Table 5 shows the contribution of the polar component to the total surface tension obtained from the geometric mean method GM for untreated and plasma treated polyurethanes. Table 5 shows that the polar term γ_{sv}^{p} in general gives a large contribution to γ_{sv}, due to the large electron donor γ_{sv}^{-} interactions. Before and after plasma treatment all samples exhibits predominant electron donor properties. Table 5 shows that the contribution of the polar component decreases after plasma treatment, except the same PPG-HPC sample. The total, disperse and polar surface tension parameters are influenced by the matrix structure of

polyurethanes possessing various soft segments. Generally, all samples possess high polar surface tension parameters, which decrease after low-pressure plasma treatment, except the PPG-HPC sample.

Polymer code	Untreated samples/ Plasma treated samples				
	γ_{sv}^{p}	γ_{sv}^{d}	γ_{sv}^{-}	γ_{sv}^{+}	γ_{sv}
PEA-HPC	57.0/24.8	2.1/16.6	55.2/23.6	12.5/4.7	59.1/41.4
PTHF-HPC	53.1/34.7	4.7/12.9	50.7/33.2	10.2/6.6	57.8/47.6
PPG-HPC	32.2/38.9	9.1/10.7	30.7/37.2	6.2/7.4	41.3/49.6

Table 4. Surface tension parameters (mN/m) for untreated and plasma treated HPC-polyurethanes according to the geometric mean method and to the acid/base method

The studied segmented cellulose polyurethanes manifest a hydrophilic character, due to cellulosic component. After HF plasma treatment the hydrophile-hydrophobe balance is changed in the sense of decreasing their hydrophilicity. This can be explained through cross-linking chemical network and by the etching effect, which modifies the rugosity and chemical composition of the surface. The exception is given by the PPG-HPC sample, which is less hydrophilic due to $-CH_3$ substituent in the soft macromolecular chain that is not favourable for polar interactions. It appears in our case that plasma induces competitive hydrophilic and hydrophobic effects and in the case of PPG-HPC sample these effects are equilibrated, such as the polar component is not changed after plasma treatment.

Polymer code	Untreated samples	Plasma treated samples
	$\gamma_{sv}^{p} / \gamma_{sv} \cdot 100$ (%)	$\gamma_{sv}^{p} / \gamma_{sv} \cdot 100$ (%)
PEA-HPC	96.5	60.0
PTHF-HPC	91.9	73.0
PPG-HPC	78.0	78.5

Table 5. Contribution of the polar component to the total surface tension obtained from the geometric mean method for untreated and plasma treated polyurethanes

The total surface tension was estimated from the structure-property relationship, according to Eqn. (6) in the following steps, (Bicerano, 1996):
1. Calculation of the zeroth-order connectivity indices $^{0}\chi$ and $^{0}\chi^{v}$ and of the first-order connectivity indices $^{1}\chi$ and $^{1}\chi^{v}$, according the values of the atomic simple connectivity indices and of the valence connectivity indices (Table 6).
2. Calculation of cohesive energy, by two methods, by applying the group contributions of Fedors, (Bicerano, 1996; Van Krevelen, 1990) and those of Van Krevelen and Hoftyzer, (Bicerano, 1996; Van Krevelen, 1990), (Table 7).
3. Calculation of the molar volume at room temperature (298 K), (Table 7).

Polymer code	$^{0}\chi$	$^{0}\chi^{v}$	$^{1}\chi$	$^{1}\chi^{v}$
PEA-HPC	179.93	139.56	121.34	32.17
PTHF-HPC	175.56	149.16	123.36	90.33
PPG-HPC	180.31	140.74	119.24	84.74

Table 6. Zeroth-order connectivity indices $^{0}\chi$ and $^{0}\chi^{v}$ and first-order connectivity indices $^{1}\chi$ and $^{1}\chi^{v}$

Polymer code	$E_{coh(1)}$, (10^{-5} J/mol)	$E_{coh(2)}$, (10^{-5} J/mol)	V (298 K), mL/mol	$\gamma_{(1)}$, mN/m	$\gamma_{(2)}$, mN/m
PEA-HPC	13.88	15.28	2812	46.89	49.95
PTHF-HPC	12.77	16.39	2613	46.61	54.76
PPG-HPC	12.25	14.75	2560	45.98	51.87

Table 7. Total surface tensions, $\gamma_{(1)}$ and $\gamma_{(2)}$, from the theoretical data calculated for cohesive energies, $E_{coh(1)}$ and $E_{coh(2)}$, and molar volume, V, for studied polyurethanes

The theoretical results are closed to the experimental values, derived from the contact angle measurements.

The hydrophobe-hydrophile balance of untreated and plasma treated polyurethanes has been evaluated by calculation of free energy of hydration, ΔG_w. The ΔG_w values were obtained from Eqn.(7), (Faibish, 2002):

$$\Delta G_w = -\gamma_{lv}(1+ \cos \theta_{water}) \qquad (7)$$

where γ_{lv} is the total surface tension of water from Table 3 and θ_{water} is contact angle of water with polyurethanes. The results are presented in Table 8.

Polymer code	Untreated samples/ plasma treated samples	
	ΔG_w (mJ/m^2)	γ_{sl} (mN/m)
PEA-HPC	-124.28/-109.2	10.6/5.0
PTHF-HPC	-124.28/-117.62	6.3/2.7
PPG-HPC	-109.2/-119.59	4.9/2.8

Table 8. Surface free energy between polyurethane and water and interfacial tensions for untreated and plasma treated samples

Generally, the literature (Faibish, 2002; Van Oss, 1994) of the field mentions that for $\Delta G_w < -113$ mJm^{-2} the polymer can be considered more hydrophilic while when $\Delta G_w > -113$ mJm^{-2} it should be considered more hydrophobic. High frequency cold plasma treatment modifies ΔG_w indicating that the surface becomes more hydrophilic in the case of PPG-HPC sample and less hydrophilic in the case of PEA-HPC and PTHF-HPC samples.

Solid-liquid interfacial tension is defined with the following relation:

$$\gamma_{sl} = (\sqrt{\gamma_{lv}^p} - \sqrt{\gamma_{sv}^p})^2 + (\sqrt{\gamma_{lv}^d} - \sqrt{\gamma_{sv}^d})^2 \qquad (8)$$

Free energy of hydration and interfacial tension are very important in that they determines the interactional force between two different media and controls the different processes: stability of the colloidal aqueous suspensions, dynamic of the molecular self-assembling, wetability of the surface, space distribution and adhesiveness. The biological and chemical processes, which take place at the level of the surface of the implant, depend on the interfacial interactions between solid and liquid (water).

(1)When the blood-biomaterial interfacial tension is high, the blood proteins will be anchored on many points on the surface, they strongly interact with the surface and thus the solid-liquid interfacial tension decreases. Consequently, the proteins change their conformation. A new interface is formed, between the protein surface and sanguine plasma.

(2)When the blood-biomaterial tension is relatively low, the force, which determines the protein adsorption, will be smaller. Conformation of the proteins initially adsorbed is similar to that found for the proteins in solution. Therefore, the interfacial tension between the protein surface and the sanguine plasma will not be high, not being an appreciable force able to determine the adsorption of sanguine components. This corresponds to a better compatibility of the biomaterial surface with blood comparing with the case (1). The surface of the biomaterial must reduce to minimum the blood-biomaterial interfacial tension such as the modification of the initially adsorbed proteins to be little. Although, apparently an interfacial tension equal to zero would be ideal for realization of the blood compatibility, however this is not desirable in view of the mechanical stability of the blood-biomaterial interface. It is generally considered that the blood-biomaterial interfacial tension should be 1-3 mN/m for a good blood - biomaterial compatibility, as well as a good mechanical stability of the interface.

The values for solid-liquid interfacial tensions are given in Table 8 for untreated and plasma treated samples. It can be observed that the interfacial tensions are in general low, and after plasma treatment become even lower. Moreover, 1 mN/m $< \gamma_{sl}$ for PTHF-HPC and PPG-HPC samples treated in plasma < 3 mN/m which is required for a good biomaterial.

An important goal in material science, biochemistry, cell biology and bioanalytics is to establish a relationship between surface morphology and surface properties of a polyurethane composite on one hand and on the other hand the interconnective interactions synthetic polymer matrix-extracellular matrix natural polymer components (Macocinschi et al., 2010a). The polyurethane was modified with natural polymers aiming the improvement of the surface and bulk properties, to confer good biomaterial qualities. The added natural extracellular matrix polymers determine an increase of the surface tension value for all the biocomposite samples in comparison with starting polyurethane. The fact that the total surface tension values experimentally evaluated for starting polyurethane (untreated and treated in plasma) are less than the theoretical one calculated on the basis of quantitative structure-properties relationships (Bicerano), which can be explained by the complex morphology of polyurethanes with microdomain segregation and network. The complex network morphology gives stability against plasma action. For all the tested biocomposites the values of the total surface tension after plasma treatment are higher, therefore the samples become more hydrophilic except one sample containing chondroitin sulfate, which suggests that chondroitin sulfate component induces further crosslinks and network formation and hence a reducing of hydrophilicity. The evaluated surface tension parameters show that the biocomposites have biomaterial qualities through their increased hydrophilicity. Moreover, plasma treatment decreases the hydration energy while the interfacial tensions fall within 1-3 mN/m, interval required for a good biomaterial. Biocompatible interfaces were constructed based on non-toxic, hydrophilic assembled macromolecular networks, as viable platforms for the affinity of biomolecules. The mechanical strength of the biocomposites decreases in comparison with starting polyurethane but it is higher than that of carotid porcine arteries.

2.2 Scanning electron microscopy

In Fig 1 are illustrated the SEM micrographs corresponding to the treated and untreated polyurethane samples. It is obvious that plasma caused a change in the surface morphology and etching effects are observed.

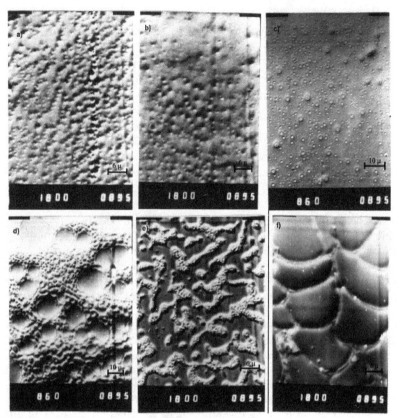

Fig. 1. SEM images of the polyurethane samples untreated and treated in HF cold plasma (PEA-HPC (a,b); PTHF-HPC (c,d); PPG-HPC (e,f))

2.3 Water sorption

The biocompatibility of the materials depend on their ability to swell in aqueous media. A high water level on the surface of the biomaterial provides a low interfacial tension with blood, thus reducing fibrinogen adsorption, cell adhesion and clot formation, (Abraham, 2002; Faibish, 2002; Van Krevelen, 1990; Wang, 2004). The results are presented in Fig 2. It is observed that the water uptake is given by the PEA-PU sample (reference polyurethane sample without hydroxypropylcellulose, PEA/MDI/EG, M_n=109.613, M_w/M_n=1.3), PEA-HPC and PTHF-HPC samples (151 %, 140 % and 167 %, respectively) and in a less extent by the PPG-HPC (92 %) due to its less polar soft segment having the lateral $-CH_3$ substituent which confer a different geometry to the polyurethane internal microporous structure, unfavorable for water uptake.

2.4 Fibrinogen adsorption

The experimental data related to the amount of adsorbed fibrinogen before and after incubating of polymers with a physiological solution of fibrinogen (3.00 mg/mL) and sanguine plasma (2.98 mg/mL) are presented in Fig 3.

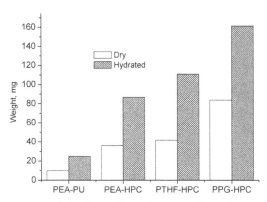
Fig. 2. Weight of polymer sample in dry and maximum hydrated state

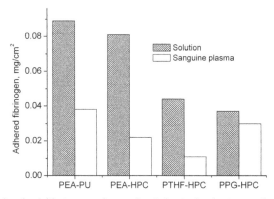
Fig. 3. Amount of adsorbed fibrinogen from physiological solution and sanguine plasma

It is observed that after incubation for 1 h at 37°C of the tested materials, no significant fluctuations of the fibrinogen concentration were recorded in comparison with the reference solution. Incubation of the samples with sanguine plasma realized in the same conditions of the incubation in solution and leaded to the results from Fig 3. Determination of the adhered fibrinogen from sanguine plasma was coupled with the determination of the prothrombin time, i.e. the time of transformation of the prothrombin in thrombin, followed by transformation of the fibrinogen in fibrin and clot formation. It is observed that the amount of fibrinogen adsorbed from sanguine plasma is less in comparison with that in solution, for all the materials except PPG-HPC, for which the differences are not significant. Also, it is remarked that prothrombin time stand in physiological normal limits (Table 9) so studied polyurethane samples did not affect the clot formation mechanisms.

The significant differences between adsorptions of the fibrinogen solution and sanguine plasma, suggest that the fibrinogen adsorption properties of the polyurethane samples, under physiological condition are affected by the concurrent affinities for other plasma proteins, which do not disturb the haemostatic mechanisms. Probably among the plasma proteins that can concur with fibrinogen is albumin, which was investigated in (Lupu et al., 2007a), and it was found that the adsorption value of a sub-physiological solution of serum albumin, (3 mg/ml) to PEA-PU materials was 0.3 ± 0.06 mg/cm^2.

Parameter	Physiological normal limits, mg/ml	Reference sample (sanguine plasma)	PEA-PU	PEA-HPC	PTHF-HPC	PPG-HPC
Prothrombin time	8.3-11.3	10.43±0.04	11.06±0.4	10.9±0.09	10.9±0.09	10.9±0.07

Table 9. Prothrombin time before and after the contact sanguine plasma with polyurethane samples

2.5 Dynamic contact angle

In Table 10 are presented the dynamic contact angle values (advancing θ_{adv},° receding θ_{rec},° and hysteresis H, %) of the film samples in contact with water.

Polymer code		θ_{adv},°	θ_{rec},°	H, %
First immersion	PEA-PU	85.31±1.11	44.21±0.53	36.31
	PEA-HPC	84.85±1.09	54.33±0.59	47.89
	PTHF-HPC	77.40±1.12	42.91±0.49	44.56
	PPG-HPC	85.63±1.08	44.81±0.51	47.67
Second immersion	PEA-PU	51.01±0.55	54.06±0.56	5.64
	PEA-HPC	52.57±0.49	43.67±0.49	16.93
	PTHF-HPC	31.61±0.41	42.26±0.45	25.20
	PPG-HPC	60.33±0.59	44.07±0.52	26.95

Table 10. Dynamic contact angles values of the film samples in contact with water

From Table 10 it is observed that at first immersion θ_{adv} are less than 90° which concludes that the polyurethane materials show hydrophilicity. These film samples have been prepared through precipitating in water making them microporous and enough hydrophile at surface through orientation of the hydrophile groups towards surface. The evaluated receding contact angles are 44-48 % less, except PEA-PU which is the polyurethane without HPC in its composition. Thus, HPC reduces the θ_{rec}. At the second immersion the hysteresis for PEA-PU is significantly less comparing with all samples having HPC. It appears that a polar material induces a low hysteresis, and the polarity of the soft segment in the polyurethane influences the results, considering additionally at second immersion the water up-take within the microporous layer disposed at surface. Advancing contact angles at second immersion are less than at first immersion, thus the material is better wet. Of all the polyurethane samples, PTHF-HPC evidences the lowest dynamic contact angles, both advancing and receding angles, at first and second immersion, which recommends it for biomedical applications.

In Table 11 are given the experimental values of the surface tension for the polyurethane solutions in DMF and DMF as solvent, by using Wilhelmy plate method. Solutions of the polyurethane sample 4g/dL in DMF were employed and DMF as a starting solvent.

Polymer solution	γ, mN/m	γ_{DMF}, mN/m	$\Delta \gamma$, %
PEA-PU	33.44±0.045	36.19±0.038	7.62
PEA-HPC	33.19±0.042	36.45±0.051	8.96
PTHF-HPC	29.26±0.035	37.36±0.044	21.69
PPG-HPC	29.51±0.031	37.34±0.041	20.96

Table 11. The experimental values of the surface tension for the polyurethane solutions in DMF and DMF as solvent

It is obvious in Table 11 that γ_{DMF} decreases only with 7.62 % by solving the PEA-PU polymer. When PEA-HPC is solved, due to cellulose derivative, OH groups and hydrogen bonding, the decrease is found to be somewhat higher. For polyether-urethanes, PTHF-HPC and PPG-HPC, there is found a much more significant $\Delta \gamma$ (21.69 %, respectively 20.96 %), and this is due to the polyether nature of the soft segment. This can be explained by the fact that when polar polymers like poly(ester urethane)s are solved in a polar solvent like DMF reduces in some extent the surface tension, while a somewhat less polar polymer like poly(ether urethane)s in the same solvent reduces more.

2.6 Dynamic mechanical thermal analysis (DMTA)

Polyurethanes are characterized by their high and heterogeneous morphology and by important intermolecular forces that determine their mechanical behavior. In such way that the forces applied to polymers in general and the deformations produced by those are not thoroughly local. Consequently the response of the polymer to the foreign solicitations is extended in a wide time interval (relaxation time), originating its peculiar viscoelastic behavior. While energy supplied to a perfectly elastic material is stored and a purely liquid dissipates it, polymeric materials dissipate a part of energy that excites to them. DMTA is a sensitive technique used to study and characterize macroscopic responses of the materials as well local internal motions. By monitoring property changes with respect to the temperature and/or frequency of oscillation, the mechanical dynamic response of the material is referred to two distinct parts: an elastic part (E', storage modulus) and a viscous component (E", loss modulus). The damping is called tan delta, loss factor or loss tangent, tan δ =E"/E'. With increasing temperature different physical states are revealed: glass, leathery, rubbery and elastic of rubbery flow, viscous flow. The glass transition is easily identified from dynamical mechanical data because of the sharp drop in storage modulus, peaks of loss dispersion modulus or tan δ. Tan δ peak may occur at higher temperatures than those given by E' drop or E" peak, because it responds to the volume fraction of the relaxing phase, its shape and height depends on the amorphous phase, being a good measure of the 'leather like' midpoint between glassy and rubbery state, (Sirear, 1997). The glass transition temperature (Tg) is often measured by DSC (Differential Scanning Calorimetry), but the DMTA technique is more sensitive and yields more easily interpreted data. This is usual as the degree of dependence is specific to the transition type. DMTA can also resolve sub-Tg transitions, like beta, gamma, and delta transitions, in many materials that the DSC technique is not sensitive enough to pick up. The magnitude of the low temperature relaxations is much smaller than that of α-relaxation considered as the glass transition. These relaxations are due to local mode (main chain) relaxations of polymer chains and rotations of terminal groups or side chains, or crankshaft motion of a few segments of the main chain.

In literature for polyurethanes based on poly(epsilon caprolactone) and 1,4-butane diisocyanate with different soft segment lengths and constant hard segment length it was evidenced by DMTA additional transitions at room temperature due to crystalline fraction of PCL while the hard segment crystallinity influence the rubber plateau, (Heijkants et al., 2005). For poly(ether urethane) networks prepared from renewable resources: epoxidized methyl-oleate polyether polyol and 1,3-propandiol by using L-lysine diisocyanate, it was found very well differentiated both from DSC and DMTA, two glass transitions for the soft segment (-17-1∘C; 9-22∘C) and respectively hard segment (35, 44∘C; 45, 58∘C) explainable through phase segregated morphology, (Lligadas et al., 2007). Three kinds of polyurethane mixed blocks (polycaprolactone glycol, polypropylene glycol, polytetramethylene glycol) and 4,4'-diphenylmethane diisocyanate extended with 1,4-butane diol were studied by DMTA and it was revealed that soft chain mobility affects the glassy state modulus. From E' and tan δ graphs the T_g values are less than -50∘C, (Mondal, 2006). Gao and Zhang, (Gao & Zang, 2001) found IPNs as a novel kind of material with complex internal friction behavior and thermal dynamic incompatibility: for their semiinterpenetrating networks of castor oil polyurethane and nitrokonjac glucomannan, glass transition temperatures ($T_g \leq 50$∘C) increased with molecular weight of the latter component which affects the storage modulus and shape and position of tan δ by changing the degree of order and motion of molecules through concentration fluctuations of molecular structural units. For graft-interpenetrating polyurethane networks and natural polymers such as nitrolignin, (Huang & Zhang, 2002), the influence of the NCO/OH molar ratio was studied by DSC and DMTA revealing T_g = -4.09-23.97∘C (DSC) and respectively (T_g= 6.3-31.1∘C) which increase with NCO/OH molar ratio through formation of three-dimensional allophanate or biuret networks. DMTA and DSC investigations on new blends of hydroxypropylcellulose and polyurethane (poly (ethylene glycol) adipate -4,4'-diphenylmethane diisocyanate-ethylene glycol) reveals glass transitions ranging between (-17-11.8∘C) by DMTA and T_m for soft segments (41.6-49.7)∘C by DSC, (Raschip et al., 2009).

Characteristic temperatures for the studied samples determined from DMTA curves (Figs. 4, 5, 6) are listed in Table 12.

Storage modulus E' is a measure of the stiffness of the material, (Zlatanic et al., 2002). It appears from the Fig. 4 that the following order of E' values at its drop (at the glass transition from glassy to leathery state) can be stated: $E'_{PEA-HPC}$> $E'_{PPG-HPC}$>E'_{PEA-PU}>$E'_{PTHF-HPC}$. This is related to the crystalline domains or physical/chemical network/entanglements which constraint molecular motions in amorphous state. For all the samples E' is less than 10^9 Pa, being generally known that when E' > 10^9 Pa the material is glassy. In addition we recall that for an amorphous linear polymer the decline of E' in the glass transition range amounts three orders of magnitude in a narrow temperature span. In particular for our samples the decline found for E' is about two orders of magnitude, polyester urethanes are expected to be stiffer, than polyether, cellulose derivative induces also stiffness to the material, while lateral methyl groups in amorphous atactic PPG provokes constraints in the mobility of the soft segment. PTHF macromolecular chain is more mobile with a low stiffness and low tan δ. The glass transition of the soft segment (SS) was determined from DMTA curves as follows: from the intersection of tangents to the E' (log E') curves from the glassy region and the transition "leathery" region, from E" (log E") peaks and tan δ peaks. T_g of the soft segment from DSC was evaluated from the second heating scan. It can be noticed that T_g from DSC are close to those from DMTA, for poly(ester urethanes) while for poly(ether urethanes) are different and this can be explained by the sensitivity of DMTA technique to the mobility of the macromolecular segment. And we referred here to the T_g values evaluated from E′ (log E′) graphs. The log E′ or log E″ vs. E′ or E″ evidence better the biphasic behaviour of the samples by revealing the melting phenomena related to the soft and hard segment. We remark slope changes on log E′ descent and the right edge of the E″ peak has a descent trend which indicate a possible overlapping of the melting of the soft phase with a glass transition of the hard phase. The broadening of the glass transition reveals large distribution of the relaxation times that implies a heterogeneous structure with a soft phase constraint by a hard phase which reduces its mobility. Poly(ether urethanes) samples evidence secondary relaxations of the soft segment, attributed to local relaxations in glassy state, which may imply few methylene groups or the motion of –CH$_3$ attached to the backbone or crankshaft motion: for PTHF-HPC sample below glass transition of the soft segment β relaxation is evidenced by tan δ graph (T_β= -51.4°C) while for PPG-HPC sample β and γ relaxations arised probably at the level of the main chain and lateral –CH$_3$ groups (E″ graph T_γ = -128°C and T_β=-81.1°C; log E″ graph T_γ = -131.6°C and T_β=-82.5°C; tan δ graph T_γ = -129.1°C and T_β= -84.4°C

Sample	Testing method	Analyzed curve	Temperature transformation, °C			
			T_g(SS)/T_α	T_m(SS)	T_m(HC)	T_β
PEA-PU	DSC		-27.0	52.8	180	-
	DMTA f=1 Hz	log E′	-23.5	65.8	157	-
		log E″	-13.4	63.6	169	-
		tan δ	15.0	78.5	-	-
PEA-HPC	DSC		-23.0	58.7	189.2	-
	DMTA f=1 Hz	log E′	-17.7	62.3	172	-
		log E″	-9.6	51.7	181	-
		tan δ	11.7	102.5	-	-
PTHF-HPC	DSC		-41.0	52.6	189.6	-
	DMTA f=1 Hz	log E′	-68.3	52.3	180	-
		log E″	-61.4	62.6	176	-
		tan δ	16	113.3	-	-51.4
PPG-HPC	DSC		-74	59.5	196	-
	DMTA f=1 Hz	log E′	-37.1	93	-	-
		log E″	-29.3	93.3	-	-82.5
		tan δ	-8.9	82.2	-	-84.4

T_m(SS) –melting point of soft segment; T_m(HS)dec. -melting point of hard segment accompanied by decomposition; T_β -secondary transitions below glass transition considered as α-transition (T_α).

Table 12. Characteristic temperatures for the studied samples determined from DMTA and DSC curves

From Fig. 6 tan δ values at glass transition peaks show that E′ is almost 5E″ for all the samples, except PTHF-HPC for which E′ is almost 10 E″. This result evidences that for all the samples the elastic modulus component is more important than the viscous modulus one.

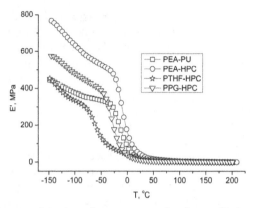

Fig. 4. Storage modulus as a function of temperature for the studied samples

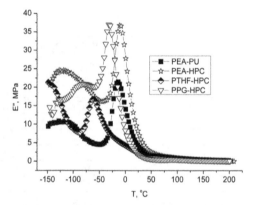

Fig. 5. Loss modulus as a function of temperature for the studied samples

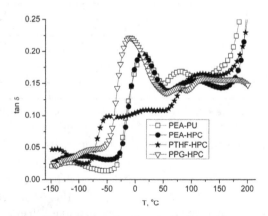

Fig. 6. Tan δ as a function of temperature for the studied samples

2.7 Mechanical properties

Biological materials have a wide range of mechanical properties matching their biological function. This is achieved via assembly of different size building block segments (soft and hard) spanning many length scales. Due to specific chemical versatility of polyurethanes, different morphologies at different length scales can be obtained and thus different physical properties which satisfy diverse clinical needs have been achieved. The modulability of mechanical properties make polyurethanes excellent candidates for applications in soft tissue engineering. Because of the strong tendency of rigid aromatic moieties to pack efficiently and the presence of hydrogen bonding between urethane and urea groups they tend to self-organize to form semi-crystalline phases within the polymer macromolecular assembly. As the elasticity of the polymers depends on their degree of crystallinity and the degree of hard segment segregation, it is clear that the selection of the diisocyanate monomer will be one of the key parameters that influence polyurethane mechanical characteristics.

The resulted tensile properties are tabulated in Table 13.

Sample	Young modulus, MPa	Elongation at break, %	Tensile strength at break, MPa	Toughness, MJ/m^3	C_1, MPa	C_2/C_1
PEA-PU	166/186	47/66	11/18	4.0/9.4	9.2/7.3	2.3/3.2
PEA-HPC	90/113	71/84	19/22	9.3/13.1	2/0.64	7.3/18.83
PTHF-HPC	70/30	72/159	14/10	7.7/11.8	3.7/0.91	4.7/3.28
PPG-HPC	75/39	53/56	15/9	5.6/3.5	4.3/0.42	4/16.03

'/' means dry/conditioned (37°C, saline water 0.9 % w/v, 24 h)

Table 13. Mechanical testing results

The stress-strain curves of the studied polyurethanes are plotted in Fig. 7 for dry film samples and in Fig. 8 the stress-strain curves of the film samples previously conditioned in warm (37 °C) saline water (NaCl, 0.9 % w/v, pH=7.4) for 24 h, then blotted with absorbent filter paper, are presented. We compare in this way, the mechanical properties of the films in dry state vs. physiological condition.

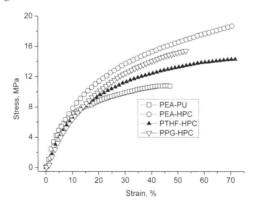

Fig. 7. Stress-strain curves of the dry polyurethanes samples

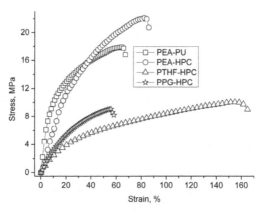

Fig. 8. Stress-strain curves of the polyurethane samples conditioned in saline water

The influence of the physical and chemical cross-links on the elastic behaviour of the polyurethanes is investigated by using Mooney-Rivlin equation, (9), for rubbers, (Sekkar et al., 2000; Spathis, 29) :

$$\sigma / (\lambda - \frac{1}{\lambda^2}) = 2C_2\lambda^{-1} + 2C_1 \qquad (9)$$

where σ is the stress, λ is the extension ratio (L/L_0) and C_1, C_2 are the Mooney-Rivlin constants.

From the stress-strain experimental data, the Mooney-Rivlin curves are plotted (Fig. 9 and Fig. 10) and the values of C_1 and C_2 are obtained (see Table 13). C_1 can be obtained by extrapolating the linear portion of the curve to $\lambda^{-1} = 0$, and C_2 from the slope of the linear portion. The elastic behaviour depends on the size and the distribution of the hard domains into the soft matrix and is more or less reflected in the deviation from the Mooney-Rivlin equation.

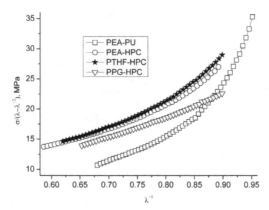

Fig. 9. Mooney-Rivlin plots of the dry polyurethanes samples

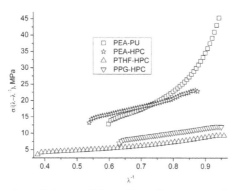

Fig. 10. Mooney-Rivlin plots of the conditioned in saline water polyurethanes samples

Moreover, the Eqn. (9) has been used to analyze the effects of different environments on the tensile behaviour of the polyurethane film samples. It has been suggested that the Mooney-Rivlin constants C_1 and C_2 are respectively associated with the network structure and the flexibility of the network. In case of dry biopolyurethanes samples it appears that polyurethane PPG-HPC shows an almost linear behaviour when comparing with PEA-PU, PEA-HPC and PTHF-HPC samples. In these block-copolymers the number of physical and chemical cross-links depends on the nature of the macrodiol used, the hydroxypropyl cellulose which generates chemical cross-links in the matrix and on the feed ratio. In the case of polyester-based matrix, PEA-HPC dry sample, the low value of C_1 evidences clearly that the more polar ester groups lead to a matrix dominated by physical cross-links. Therefore, C_1 is different for HPC polyether- and polyester-urethanes (C_1 is lower for polyester- than polyether-urethanes). At higher strains a disrupting of the physical cross-links is taking place and C_1 becomes lower. C_2 is not so sensitive relative to the nature of soft segment but shows the presence of both physical and chemical cross-links specific to a polyurethane network. The effect of conditioning in saline water (0.9 % w/v, 24 h, 37°C) is clearly revealed by C_1 which is found lower for all the samples evidencing that after conditioning the physical network is affected as well as its flexibility.

The toughness representing the energy absorbed before the sample breaks is higher as expected for PEA-HPC sample than for the PPG-HPC and PTHF-HPC samples, both for dry and conditioned samples. Moreover it can be noticed that the cellulose derivative makes that this absorbed energy to be much higher in case of dry samples. Hydration of samples leads to the increase of toughness except PPG-HPC sample due to amorphous and atactic soft segment structure.

Hydration of the semicrystalline, more or less ordered structures like polyurethanes, have as main result the disrupting of the physical bonding, and the plasticizing of the biopolyurethane matrix, affecting strain behaviour. Water may penetrate within interstitials of the microporous structure favoring biological interactions. The heat is also important in softening the material acting upon the soft segment and physical hydrogen bonding. From Table 13 one can notice that for polyether-urethanes (PTHF-HPC, PPG-HPC), Young modulus decreases after conditioning in saline water at 37°C for 24 h, the material achieve more flexibility, which may bring as advantage for realization of biopolyurethane tissular structures, such heart valves and leather grafts.

Hydrogen bonding is known as an important driving force for the phase separation of a hard segment from a soft-segment matrix consisting of polyether or polyester polyol. The separated hard segment acts as physical cross-links and filler particles for the soft-segment matrix. Microphase separation of segmented polyurethanes is probably the single most influential characteristic of these materials. The degree of phase separation plays a key role in determining mechanical properties and blood compatibility, (Yoon et al., 2005). The heterogeneous morphology of polyurethanes will determine the surface composition exposed to a polar environment (water or blood) or to a nonpolar environment (air or vacuum). The surface segregation phenomenon reflects the difference in surface energy between the polar and nonpolar components, (Lupu et al., 2007b). The surface composition constitutes a crucial parameter for a biomedical material in contact with blood. The mobility of polymer chains coupled with environmental changes can lead to surface composition and properties that are time-dependent and dependent on the contacting medium, temperature, that the polymer experiences. It is, however, an unresolved question as to whether an air-stored polyurethane surface indeed adapts to the aqueous environment on biomedical usage. As the polyurethane biomaterial is placed into contact with a physiological medium, such as blood or tissue, its surface layers will undergo motions in order to accommodate the new interfacial situation. In contact with aqueous environments, it is obviously favorable for hydrophilic constituents of the polymer to become enriched at the interface, (Vermette & Griesser, 2001).

For crosslinked blends of Pellethene and multiblock polyurethanes containing phospholipids, (Yoo & Kim, 2005), it was found for elastic modulus values ranged between 21-47 MPa, whereas for biomaterials based on cross-linked blends of Pellethene and multiblock polyurethanes containing poly(ethylene oxide) it was found for elastic modulus values ranged between 85-246 MPa, (Yoo & Kim, 2004).

2.8 Platelet adhesion

In vitro platelet adhesion experiments were conducted to evaluate the preliminary blood compatibility. It is very well known that the surface properties particularly platelet adhesion of the biomaterials is very important with respect to their haemocompatibility, especially when they are used as cardiovascular devices, (Park et al., 1998).

It is well known that when blood is in contact with a synthetic material, firstly the latter one adsorbs onto its surface blood plasma proteins, and secondly attract and activate the thrombocytes. In function of the type of the adsorbed plasma proteins (fibrinogen or albumins) this phenomenon can exert in a more or less extent. In case of the preferentially fibrinogen adsorption (important protein in endogenous haemostasis), platelet adhesion is increased followed by thrombus activation and clot formation. In case of albumin absorbtion platelet adhesion is diminished which confer to the surface a thromboresistant character, (Bajpai, 2005; Wang et al., 2004).

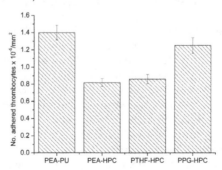

Fig. 11. Platelet adhesion on the studied biopolyurethanes film surfaces

In Fig. 11 platelet adhesion corresponding to the studied polyurethane samples is given. It can be noticed that when comparing PEA-PU with PEA-HPC the no. of adhered thrombocytes is much lower for PEA-HPC and this is due to the cellulose derivative which confers improved biomaterial qualities. Among the PPG-HPC, PTHF-HPC and PEA-HPC samples the most promising thromboresistant ones are PTHF-HPC and PEA-HPC. In our previous paper, the fibrinogen adsorption tests, (Macocinschi et al., 2009a), revealed that PTHF-HPC and PEA-HPC polyurethanes proved to be indicated for thromboresistant devices, and the polyether-urethane PTHF-HPC proved to have more relevant haemocompatible material qualities.

From literature for Biospan polyurethane, (Korematsu et al., 1999), the values for adhered thrombocytes found are 101500 adhered thrombocytes per mm^2. The number of adhered platelet is sensitive also to the soft segment length in the case of fluorinated polyurethanes, (Wang & Wei, 2005). The presence of cellulose in segmented polyurethane matrix indeed inhibits platelet adhesion, (Hanada et al., 2001). For PU/hydrophilic poly(ethylene glycol)diacrylate IPNs, platelet adhesion is suppressed by microseparated IPN structure, (Yoon et al., 2005). For poly(carbonate urethane)s with various degree of nanophase segregation the number of platelet adhered found was 5.33 x 10^6 up to 10.67 x 10^6 (for Pellethane is 8x10^6), (Hsu & Kao, 2005).

3. Conclusion

The effect of the chemical structure of some cellulose derivative based segmented biopolyurethanes and polyurethane biocomposites containing extracellular matrix components on surface properties (static and dynamic contact angle measurements), on dynamical mechanical thermal analysis, mechanical properties was investigated. Specific biological tests were performed (platelet adhesion, protein adsorption tests of fibrinogen) and the results obtained recommend them as biomaterials with enhanced biocompatibility for biomedical applications. Thus, the proper understanding of blood compatibility and the development of prosthetic materials require the investigations of both natural and synthetic materials. Extensive in vitro, ex vivo, and in vivo testing in suitable animals and related differences between the properties of blood components of animals and humans are essential. Future research should be directed in order to clarify the influence of anatomical site of implantation on the behaviour of the implant, with a goal towards the development of site specific implants.

4. Acknowledgment

This work was supported by CNCS-UEFISCDI, Research Projects, Code: PN-II-ID-PCE-988/2008, contract no. 751/2009.

5. References

Abraham, G.A., De Queiroz, A.A.A. & Roman, J.S. (2002). Immobilization of a Nonsteroidal Antiinflammatory Drug onto Commercial Segmented Polyurethane Surface to Improve Haemocompatibility Properties. *Biomaterials*, Vol. 23, No. 7, (April 2002), pp. 1625-1638, ISSN 0142-9612

Bajpai, A.K. (2005). Blood Protein Adsorption onto a Polymeric Biomaterial of Polyethylene glycol and Poly[(2-hydroxyethylmethacrylate)-co-acrylonitrile] and Evaluation of In Vitro Blood Compatibility. *Polymer International*, Vol. 54, No. 2, (February 2005), pp. 304-315, ISSN 1097-0126

Baumgartner, J.N., Yang, C.Z. & Cooper, S.L. (1997). Physical Property Analysis and Bacterial Addhesion on a Series of Phosphonated Polyurethanes. *Biomaterials*, Vol. 18, No. 12, (May 1998), pp. 831-837, ISSN 0142-9612

Bicerano, J. (1996). Prediction of the Properties of Polymers from Their Structures. *Journal of Macromolecular Science-Reviews in Macromolecular Chemistry and Physics*, Vol. 36, No. 1, (February 1996), pp. 161-196, ISSN 0736-6574

Desai, S., Thakore, I.M., Sarawade, B.D. & Dewi, S. (2000). Effect of polyols and diisocyanates on thermo-mechanical and morphological properties of polyurethanes. *European Polymer Journal*, Vol. 36, No. 4, (April 2000), pp. 711-725, ISSN 0014-3057

Durairaj, R.B. (2001). HER Materials for Polyurethane Applications, In: *Advances in Urethane Science and Technology*, Klempner, D. & Frisch, K.C. (Eds.), pp. 369-419, Rapra Technology Ltd., ISBN 1-85957-275-8, Shawbury, UK

Erbil, H.Y. (1997). Surface Tension of Polymers, In: *Handbook of Surface and Colloid Chemistry*, Birdi, K.S. (Ed.), pp. 265-312, CRC Press, ISBN 0-8493-9459-7, Boca Raton, USA

Faibish, R.S., Yoshida, W. & Cohen, Y. (2002). Contact Angle Study on Polymer-Grafted Silicon Wafers. *Journal of Colloid and Interface Science*, Vol. 256, No. 2, (December 2002), pp. 341-350, ISSN 0021-9797

Gao, S. & Zhang, L. (2001). Molecular Weight Effects on Properties of Polyurethane/Nitrokonjac Glucomannan Semiinterpenetrating Polymer Netwoks. *Macromolecules*, Vol. 34, No. 7, (March 2001), pp. 2202-2207, ISSN 0024-9297

Grundke, K. (2005). Surface-energetic Properties of Polymers in Controlled Architecture, In: *Molecular Interfacial Phenomena of Polymers and Biopolymers*, Chen, P. (Ed.), pp. 323-374, Woodhead Publishing Ltd., ISBN-13 978-1-85573-928-4, Cambridge, England

Hanada, T., Li, Y.J. & Nakaya, T. (2001). Synthesis and Haemocompatibilities of Cellulose-Containing Segmented Polyurethanes. *Macromolecular Chemistry and Physics*, Vol. 202, No. 1, (January 2001), pp. 97-104, ISSN 1022-1352

Heijkants, R.G.J.C., Van Calck, R.V., Van Tienen, T.G., De Groot, J.H., Buma, P., Pennings, A.J., Veth, R.P.H. & Schouten, A.J. (2005). Uncatalyzed Synthesis, Thermal and Mechanical Properties of Polyurethanes based on Poly(epsilon-caprolactone) and 1,4-Butane Diisocyanate with Uniform Hard Segment. *Biomaterials*, Vol. 26, No. 20, (July 2005), pp. 4219-4228, ISSN 0142-9612

Hsu, S. & Kao, Y.C. (2005). Biocompatibility of Poly(carbonate urethane)s with Various Degrees of Nanophase Separation. *Macromolecular Bioscience*, Vol. 5, No. 3, (March 2005), pp. 246-253, ISSN 1616-5187

Huang, J. & Zhang, L. (2002). Effect of NCO/OH Molar Ratio on Structure and Properties of Graft-Interpenetrating Polymer Networks from Polyurethane and Nitrolignin. *Polymer*, Vol. 43, No. 8, (April 2002), pp. 2287-2294, ISSN 0032-3861

Hung, H.S., Wu, C.C., Chien, S. & Hsu, S.H. (2009). The Behavior of Endothelial Cells on Polyurethane Nanocomposite and the Associated Signaling Pathways. *Biomaterials*,Vol. 30, No. 8, (March 2009), pp. 1502-1511, ISSN 0142-9612

Kalble, D.H. (1969). Peel Adhesion: Influence of Surface Energies and Adhesive Rheology. *The Journal of Adhesion*, Vol. 1, No. 2, (February 1969), pp. 102-123, ISSN 0021-8464

Korematsu, A., Tomita, T., Kuriyama, S., Hanada, T., Sakamoto, S. & Nakaya, T. (1999). Synthesis and Blood Compatibilities of Novel Segmented Polyurethanes Grafted Phospholipids Analogous Vinyl Monomers and Polyfunctional Monomers. *Acta Polymerica*, Vol. 50, No. 10, (October 1999), pp. 363-372, ISSN 0323-7648

Lamba, N.M.K., Woodhouse, K.A. & Cooper, S.L. (1997). *Polyurethanes in Biomedical Applications*, CRC Press, ISBN-10 0849345170, New York

Lelah, M.D. & Cooper, S.L. (1986). *Polyurethanes in medicine*, CRC Press, ISBN-10 0849363071, Boca Raton, FL

Lligadas, G., Ronda, J.C, Galia, M. & Cadiz, V. (2007). Poly(ether urethane) Networks from Renewable Resources as Candidate Biomaterials: Synthesis and Characterization. *Biomacromolecules*, Vol. 8, No. 2, (February 2007), pp. 686-692, ISSN 1525-7797

Lupu, M., Butnaru, M., Macocinschi, D., Oprean, O.Z., Dimitriu, C., Bredetean, O., Zagnat, M. & Ioan, S. (2007a). Surface Properties of Segmented Poly(ester urethane)s and Evaluation of In Vitro Blood Compatibility and In Vivo Biocompatibility. *Journal of Optoelectronics and Advanced Materials*, Vol. 9, No. 11, (November 2007), pp. 3474-3478, ISSN 1454-4164

Lupu, M., Macocinschi, D., Ioanid, G., Butnaru, M. & Ioan, S. (2007b). Surface Tension of Poly(ester urethane)s and Poly(ether urethane)s. *Polymer International*, Vol. 56, No. 3, (March 2007), pp. 389-398, ISSN 1097-0126

Macocinschi, D., Filip, D. & Vlad, S. (2008). New Polyurethane Materials from Renewable Resources: Synthesis and Characterization. *e-Polymers*, no.062, (May 2008), ISSN 1618-7229

Macocinschi, D., Filip, D., Butnaru, M. & Dimitriu, C.D. (2009a). Surface Characterization of Biopolyurethanes Based on Cellulose Derivatives. *Journal of Materials Science: Materials in Medicine*, Vol. 20, No. 3 (March 2009), pp. 775-783, ISSN 0957-4530

Macocinschi, D., Filip, D., Vlad, S., Cristea, M. & Butnaru, M. (2009b). Segmented Biopolyurethanes for Medical Applications. *Journal of Materials Science: Materials in Medicine*, Vol. 20, No. 8, (August, 2009), pp. 1659-1668, ISSN 0957-4530

Macocinschi, D., Filip, D. & Vlad, S. (2010a). Surface and Mechanical Properties of Some New Biopolyurethane Composites. *Polymer Composites*, Vol. 31, No. 11, (November 2010), pp. 1956-1964, ISSN 1548-0569

Macocinschi, D., Tanase, C., Filip, D., Vlad, S. & Oprea, A. (2010b). Study of the Relationship Between New Polyurethane Composites for Biomedical Applications and Fungal Contamination. *Materiale Plastice*, Vol. 47, No. 3, (September 2010), pp. 286-291, ISSN 0025-5289

Macocinschi, D., Filip, D., Vlad, S., Cristea, M., Musteata, V.E. & Ibanescu, S. (2011). Thermal, Dynamic Mechanical, and Dielectric Analyses of Some Polyurethane Biocomposites. *Journal of Biomaterials Applications*, DOI 10.1177/0885328210394468, ISSN 0885-3282

Moldovan, L., Craciunescu, O., Zarnescu, O., Macocinschi, D. & Bojin, D. (2008). Preparation and Characterization of New Biocompatibilized Polymeric Materials for Medical Use. *Journal of Optoelectronics and Advanced Materials*, Vol. 10, No. 4, (April 2008), pp. 942-947, ISSN 1454-4164

Mondal, S. & Hu, J.L. (2006). Structural Characterization and Mass Transfer Properties of Polyurethane Block Copolymer: Influence of Mixed Soft Segment Block and Crystal Melting Temperature. *Polymer International*, Vol. 55, No. 9, (September 2006), pp. 1013-1020, ISSN 1097-0126

Musteata, V.E., Filip, D., Vlad, S. & Macocinschi, D (2010). Dielectric Relaxation of Polyurethane Biocomposites. *Optoelectronics and Advanced Materials-Rapid Communications*, Vol. 4, No. 8, (August 2010), pp. 1187-1192, ISSN 1842-6573

Norde, W. (2006). Surface Modifications to Influence Adhesion of Biological Cells and Adsorption of Globular Proteins, In: *Surface Chemistry in Biomedical and Environmental Science*, Blitz, J.P. & Gun'Ko, V.M. (Eds.), pp. 159-176, Springer, ISBN-10 1-4020-4740-1, Dordrecht, Netherlands

Ohya, Y. (2002). Hydrogen Bonds, In: *Supramolecular Design for Biological Applications*, Yui, N. (Ed.), pp. 33-72, CRC Press, ISBN 0-8493-0965-4, Boca Raton, Florida

Owens, D.K. & Wendt, R.C. (1969). Estimation of Surface Free Energy of Polymers. *Journal of Applied Polymer Science*, Vol. 13, No. 8, (March 2003), pp. 1741-1747, ISSN 1097-4628

Ozdemir, Y., Hasirici, N. & Serbetci, K. (2002). Oxygen Plasma Modification of Polyurethane Membranes. *Journal of Materials Science: Materials in Medicine*, Vol. 13, No. 12, (December 2002), pp. 1147-1152, ISSN 0957-4530.

Park, K.D., Okano, T., Nojiri, C. & Kim, S.W. (1988). Heparin Immobilization onto Segmented Polyurethaneurea Surfaces-Effect of Hydrophilic Spacers. *Journal of Biomedical Materials Research*, Vol. 22, No. 11, (November 1988), pp. 977-992, ISSN 1549-3296

Pilkey, W.D. (2005). *Formulas for Stress, Strain, and Structural Matrices*, Second edition, Wiley & sons. Inc., ISBN 0-471-03221-2, Hoboken, New Jersey

Plank, H., Syre, I., Dauner, M. & Egberg, G. (Eds.). (1987). *Polyurethane in Biomedical Engineering: II. Progess in Biomedical Engineering 3*, Elsevier Science, ISBN 0-44-442759-7, Amsterdam

Rabel, W. (1977). Flussigkeitsgrenzflachen in Theorie und Anwendungstechnik. *Physicalische Blatter*, Vol. 33, pp. 151-161, ISSN 0031-9279

Ramis, X., Cadenato, A., Morancho, J.M. & Salla, J.M. (2001). Polyurethane-Unsaturated Polyester Interpenetrating Polymer Networks: Thermal and Dynamic Mechanical Thermal Behaviour. *Polymer*, Vol. 42, No. 23, (November 2001), pp. 9469-9479, ISSN 0032-3861

Rankl, M., Laib, S. & Seeger, S. (2003). Surface Tension Properties of Surface-Coatings for Application in Biodiagnostics Determined by Contact Angle Measurements. *Colloids and Surfaces B: Biointerfaces*, Vol. 30, No. 3, (July 2003), pp. 177-186, ISSN 0927-7765

Raschip, I.E., Vasile, C. & Macocinschi, D. (2009). Compatibility and Biocompatibility Study of New HPC/PU Blends. *Polymer International*, Vol. 58, No. 1, (January 2009), pp. 4-16, ISSN 1097-0126

Reddy, T.T., Kano,A., Maruyama, A., Hadano, M. & Takahara, A. (2008). Thermosensitive Transparent Semi-Interpenetrating Polymer Networks for Wound Dressing and Cell Adhesion Control. *Biomacromolecules*, Vol. 9, No. 4, (April 2008), pp. 1313-1321, ISSN 1525-7797

Sekkar, V., Bhagawan, S.S., Prabhakaran, N., Rama, R.M. & Ninan, K.N.(2000). Polyurethanes Based on Hydroxyl Terminated Polybutadiene: Modeling of

Network Parameters and Correlation with Mechanical Properties. *Polymer*, Vol. 41, No. 18, (August 2000), pp. 6773-6786, ISSN 0032-3861

Siegwart, R. (2001). Indirect Manipulation of a Sphere on a Flat Disk Using Force Information. *International Journal of Advanced Robotic Systems*, Vol. 6, No. 4, (December 2009), pp. 12-16, ISSN 1729-8806

Sirear, A.K. (1997). Elastomers. In: *Thermal Characterization of Polymeric Materials*. Turi, E.A. (Ed.), Vol. 1, Academic Press, pp. 970-1025, ISBN 0-12-703782-9

Spathis, G.D. (1991). Polyurethane Elastomers Studied by the Mooney Rivlin Equation for Rubbers. *Journal of Applied Polymer Science*, Vol. 43, No. 3, (August 1991), pp. 613-620, ISSN 1097-4628

Strom, G., Fredriksson, M. & Stenius, P. (1987). Contact Angles, Work of adhesion, and Interfacial Tensions at a Dissolving Hydrocarbon Surface. *Journal of Colloid and Interface Science*, Vol. 119, No. 2, (October 1987), pp. 352-361, ISSN 0021-9797

Thomson, T. (2005). *Polyurethanes as Specialty Chemicals. Principles and Applications*, CRC Press, ISBN 0-8493-1857-2, Boca Raton, Florida

Van Krevelen, D.W. (1990). *Properties of Polymers; Their Correlation with Chemical Structure, their Numerical Estimation and Prediction from Additive Group Contributions*, Elsevier, ISBN 0-444-88160-3, Amsterdam

Van Oss, C.J., Good, R.J. & Chaudhury, M.K. (1988a). Additive and Nonadditive Surface Tension Components and the Interpretation of Contact Angles. *Langmuir*, Vol. 4, No. 4, (July 1988), pp. 884-891, ISSN 0743-7463

Van Oss, C.J., Ju, M.K., Chaudhury, M.K. & Good, R.J. (1988b), Interfacial Lifshitz-van der Waals and Polar Interactions in Macroscopic Systems. *Chemical Reviews*, Vol. 88, No. 6, (September 1988), pp. 927-941, ISSN 0009-2665

Van Oss, C.J., Ju, L., Chaudhury, M.K. & Good, R.J. (1989). Estimation of the Polar Parameters of the Surface Tension of Liquids by Contact Angle Measurements on Gels. *Journal of Colloid and Interface Science*, Vol. 128, No. 2, (March 1989), pp. 313-319, ISSN 0021-9797

Van Oss, C.J. (1994). *Interfacial Forces in Aqueous Media*, Marcel Dekker, ISBN 0-8274-9168-1, New York

Vermette, P., Griesser, H.J., Laroche, G. & Guidoin, R. (2001). *Biomedical Applications of Polyurethanes. Tissue Engineering Intelligence Unit 6*, Eurekah.com, ISBN 1-58706-023-X, Texas, USA

Wang, L.F. & Wei, Y.H. (2005). Effect of Soft Segment Length on Properties of Fluorinated Polyurethanes. *Colloids and Surfaces B: Biointerfaces*, Vol. 41, No. 4, (April 2005), pp. 249-255, ISSN 0927-7765

Wang, Y.X., Robertson, J.L., Spillman Jr., W.B. & Claus, R.O. (2004). Effects of the Chemical Structure and the Surface Properties of Polymeric Biomaterials on Their Biocompatibility. *Pharmaceutical Research*, Vol. 21, No. 8, (August 2004), pp. 1362-1373, ISSN 0724-8741

Wu, Z.Q., Chen, H., Huang, H., Zhao, T., Liu, X., Li, D. & Yu, Q. (2009). A Facile Approach to Modify Polyurethane Surfaces for Biomaterial Applications. *Macromolecular Bioscience*, Vol. 9, No. 12, (December 2009), pp. 1165-1168, ISSN 1616-5195

Yoo, H.J. & Kim, H.D. (2004). Properties of Crosslinked Blends of Pellethene and Multiblock Polyurethane Containing Poly(ethylene oxide) for Biomaterials. *Journal of Applied Polymer Science*, Vol. 91, No. 4, (February 2004), pp. 2348-2357, ISSN 1097-4628

Yoo, H.J. & Kim, H.D. (2005). Characteristics of Crosslinked Blends of Pellethene and Multiblock Polyurethanes Containing Phospholipids. *Biomaterials*, Vol. 26, No. 16, (June 2005), pp. 2877-2886, ISSN 0142-9612

Yoon, S.S., Kim, J.H. & Kim, S.C. (2005). Synthesis of Biodegradable PU/PEGDA IPNs Having Micro-Separated Morphology for Enhanced Blood Compatibility. *Polymer Bulletin*, Vol. 53, No. 5-6, (March 2005), pp. 339-347, ISSN 0170-0839

Zlatanic, A., Petrovic, Z.S. & Dusek, K. (2002). Structure and Properties of Triolein-Based Polyurethane Networks. *Biomacromolecules*, Vol. 3, No. 5, (September 2002), pp. 1048-1056, ISSN 1525-7797

5

Collagen-Based Drug Delivery Systems for Tissue Engineering

Mădălina Georgiana Albu[1], Irina Titorencu[2] and Mihaela Violeta Ghica[3]
[1]INCDTP – Leather and Footwear Research Institute, Bucharest
[2]Institute of Cellular Biology and Pathology "Nicolae Simionescu", Bucharest
[3]Carol Davila University of Medicine and Pharmacy, Faculty of Pharmacy, Bucharest
Romania

1. Introduction

Biomaterials are considered those natural or artificial materials that can be used for any period of time, as a whole or as part of a system which treats, augments or replaces a tissue, organ or function of the human or animal body (Williams, 1999). In medicine a wide range of biomaterials based on metals, ceramics, synthetic polymers, biopolymers, etc. is used. Among biopolymers, collagen represents one of the most used biomaterials due to its excellent biocompatibility, biodegradability and weak antigenecity, well-established structure, biologic characteristics and to the way it interacts with the body, the latter recognizing it as one of its constituents and not as an unknown material (Friess, 1998; Lee et al., 2001). Irrespective of the progress in the field of biomaterials based on synthetic polymers, collagen remains one of the most important natural biomaterials for connective tissue prosthetic in which it is the main protein. Due to its excellent properties collagen can be processed in different biomaterials used as burn/wound dressings, osteogenic and bone filling materials, antithrombogenic surfaces, collagen shields in ophthalmology, being also used for tissue engineering including skin replacement, bone substitutes, and artificial blood vessels and valves. Biomaterials based on type I fibrillar collagen such as medical devices, artificial implants, drug carriers for controlled release and scaffolds for tissue regeneration have an important role in medicine, being widely used at present (Healy et al., 1999; Hubell, 1999; Wang et al., 2004). In this chapter, we attempted to summarize some of the recent developments in the application of collagen as biomaterial in drug delivery systems and tissue engineering field.

2. Collagen-based biomaterials

Collagen is the main fibrous protein constituent in skin, tendons, ligaments, cornea etc. It has been extensively isolated from various animals, including bovine (Renou et al., 2004; Doillon, 1992), porcine (Smith et al., 2000; Lin et al., 2011; Parker et al., 2006), equine (Angele et al., 2004), ovine (Edwards et al., 1992), shark, frog, bird (Limpisophon et al., 2009) and from marine origin such as: catfish (Singh et al., 2011), silver carp (Rodziewicz-Motowidło et al., 2008), marine sponge (Swatschek et al., 2002), jumbo squid (Uriarte-Montoya et al., 2010), paper nautilus (Nagai & Suzuki, 2002), tilapia fish-scale (Chen et al., 2011), red fish (Wang et al., 2008). Among these types of sources the most used has been bovine hide. Although to date 29 different types of collagen have been identified (Albu, 2011), type I collagen is the most abundant and still the best studied. This work is focused on biomaterials based on type I collagen of bovine origin. Type I collagen consists of 20 amino acids, arranged in characteristic sequences which form a unique conformational structure of triple helix (Trandafir et al., 2007). Hydroxyproline is characteristic only for collagen and it confers stability for collagen, especially by intramolecular hydrogen bonds. The collagen structure is very complex, being organised in four levels, named primary, secondary, tertiary and quaternary structure. Depending on the process of collagen extraction, the basic forms of collagen are organized on structural level.

2.1 Process of collagen extraction
To obtain extracts of type I fibrillar collagen, fresh skin or skin technological waste from leather industry can be used as raw materials (Trandafir et al., 2007), extraction being performed from dermis. To minimize the exogenous degradation the skin has to be ready for immediate extraction. Yield of good extraction is obtained from skin of young animals (preferably younger than two years) due to weaker crosslinked collagen.
Figure 1 schematically shows the obtaining of collagen in different forms by the currently used technologies at Collagen Department of Leather and Footwear Research Institute, Bucharest, Romania.
As figure 1 shows, the bovine hide was used as raw material. After removal of hair and fat by chemical, enzymatic or mechanical process, the obtained dermis could undergo different treatment and *soluble* or *insoluble collagen* is obtained.

2.1.1 Process of extraction for soluble collagen
Depending on structural level the solubilised collagen extracts can be denatured (when 90% of molecules are in denatured state) or un-denatured (when 70% of molecules keep their triple helical structure) (Trandafir et al., 2007).
The process for obtaining of *denatured collagen* took place at high temperature, pressure or concentrated chemical (acid or alkali) or enzymatic agents. Following these critical conditions the collagen is solubilised until secondary or primary level of structure and *gelatine* or *partial* (polypeptide) and *total* (amino acids) *hydrolisates* are obtained.
The *undenatured collagen* can be isolated and purified by two technologies, depending on the desired structural level (Li, 2003): molecular and fibrillar. They allow the extraction of type I collagen from bovine hide in aqueous medium while maintaining the triple helical structure of molecules, of microfibrils and fibres respectively (Wallace & Rosenblatt, 2003).
Isolation and purification of collagen molecules from collagenic tissues can be performed using a proteolytic enzyme such as pepsin, which produces cleavage of telopeptides - places responsible for collagen crosslinking. Removing them makes the collagen molecules and small aggregates (protofibrils) soluble in aqueous solutions of weak acid or neutral salts.
Extraction of collagen soluble in neutral salts. Studies on the extraction of soluble collagen with neutral salt solutions were performed with 0.15 to 0.20 M sodium chloride at 5°C for 1-2 days (Fielding, 1976). Yield of this technology is low and the most collagenic tissues extracted with salts contain small quantities of collagen or no collagen at all.

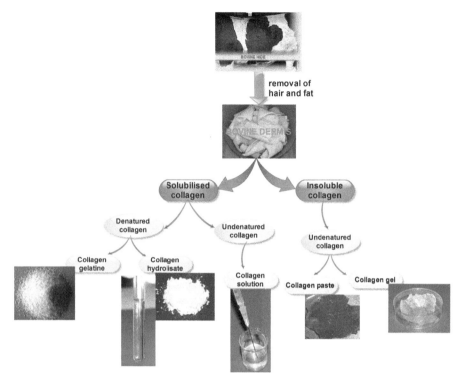

Fig. 1. Basic forms of collagen

Extraction of acid soluble collagen. Dilute acids as acetic, hydrochloric or citrate buffer solution with pH 2-3 are more effective for extraction of molecular collagen than neutral salt solutions. Type aldiminic intermolecular bonds are disassociated from dilute acids and by exerting forces of repulsion that occur between the same charges on the triple helix, causing swelling of fibril structure (Trelstad, 1982). The diluted acids do not dissociate keto-imine intermolecular bonds. For this reason collagen from tissues with high percentage of such bonds, such as bone, cartilage or tissue of aged animals is extracted in smaller quantities in dilute acids.

To obtain soluble collagen with diluted acids tissue is ground cold, wash with neutral salt to remove soluble proteins and polysaccharides, then collagen is extracted with acid solutions (Bazin & Delaumay, 1976). Thus about 2% of collagen can be extracted with salts or diluted acid solutions.

Enzymatic extraction is more advantageous, collagen triple helix being relatively resistant to proteases such as pronase, ficin, pepsin or chemotripsin at about 20°C (Piez, 1984). The efficacy of enzymatic treatment arises from selective cleavage in the terminal non-helical regions monomer and higher molecular weight covalently linked aggregates, depending on the source and method of preparation. Thus, telopeptidic ends are removed, but in appropriate conditions the triple helices remain intact. Solubilised collagen is purified by salt precipitation, adjusting pH at the isoelectric value or at temperature of 37°C (Bazin & Delaumay, 1976). Collagen extracted with pepsin generally contains higher proportions of intact molecules extracted with salts or acids.

2.1.2 Process of extraction for insoluble collagen

Collagen extraction by alkaline and enzymatic treatments. Alkaline pretreatment destroys covalent bonds resistant to acids. Collagen interaction with alkali shows the presence of certain specificities, hydrogen bonds being more sensitive to alkali. Degradation of the structure is more intense and irreversible if treatment is progressing on helicoidal structure (collagen → gelatin transition, alkaline hydrolysis).

Breaking of hydrogen bonds occurs by replacing the hydrogen atom from carboxyl groups with metal which is unable to form hydrogen bonds. Collagen can be extracted by treating the dermis with 5-10% sodium hydroxide and 1 M sodium sulphate at 20-25°C for 48 hours (Cioca, 1981; Trandafir et al., 2007). Thus, fats associated with insoluble collagen are saponified, the telopeptidic non-helical regions are removed, collagen fibers and fibrils are peptized. Size of resulted fragments of collagen depends on the time and concentration of alkali treatment (Roreger, 1995). The presence of sodium sulfate solution controlled the swelling of collagen structure, protecting the triple-helical native conformation. Alkaline treatment is followed by an acid one, which leads to total solubilization of collagen in undenatured state from the dermis of mature animals.

Thus technologies of molecular and fibrilar extraction are enabled to extract type I collagen from bovine hide in an aqueous medium keeping triple helical structure of molecules, microfibrils and fibrils (Wallace & Rosenblatt, 2003).

2.2 Obtaining of collagen-based biomaterials

Obtaining of collagen-based biomaterials starts from undenatured collagen extracts – gels and solutions – which are processed by cross-linking, free drying, lyophilization, elecrospinning, mineralisation or their combinations. To maintain the triple helix conformation of molecules the conditioning processes must use temperatures not higher than 30°C (Albu et al., 2010a). Extracted as aqueous solution or gel, type I collagen can be processed in different forms such as hydrogels, membranes, matrices (spongious), fibers, tubes (Fig. 2) that have an important role in medicine today. Figure 2 shows some collagen-based biomaterials obtained at our Collagen Department.

Among the variety of collagen-based biomaterials, only the basic morphostructural ones will be presented: hydrogels, membranes, matrices, and composites obtained from undenatured collagen.

Collagen hydrogels are biomaterials in the form of tridimensional networks of hydrofil polymeric chains obtained by physical or chemical cross-linking of gels. Chemical cross-linking consists in collagen reaction with aldehydes, diisocyanates, carboimides, acyl-azide, polyepoxydes and polyphenolic compounds which lead to the formation of ionic or covalent bonds between molecules and fibrils (Albu, 2011). Physical cross-linking includes the drying by heating or exposure at UV, gamma or beta irradiations. Their mechanical and biological properties are controllable and superior to the gels from which they were obtained.

The hydrogels have the capacity of hydration through soaking or swelling with water or biological fluids; hydrogels with a solid laminar colloidal or solid sphero-colloidal colloidal frame are formed, linked by means of secondary valences, where water is included by swelling. One of the exclusive properties of hydrogels is their ability to maintain the shape during and after soaking, due to the isotropic soaking. Also the mechanical properties of the collagen hydrogels are very important for the pharmaceutical applications, the modification of the cross-linking degree leading to the desired mechanical properties. The spreading ability of the different size molecules in and from hydrogels serves for their utilization as drug release systems. The development and utilization of collagen hydrogels in therapeutics is supported by some advantages contributing to patients compliance and product efficiency. Thus, the hydrogels are easy to apply, have high bioadhesion, acceptable viscosity, compatibility with numerous drugs (Albu & Leca, 2005; Satish et al., 2006; Raub et al., 2007).

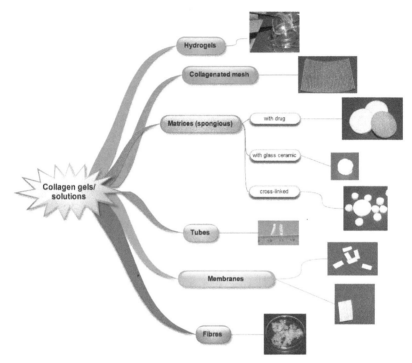

Fig. 2. Collagen-based biomaterials

Collagen membranes/films are obtained by free drying of collagen solution/gel in special oven with controllable humidity and temperature (not higher than 25°C) during 48-72 hours. These conditions allow the collagen molecules from gels to be structured and to form intermolecular bonds without any cross-linking agent. They have dense and microporous structure (Li et al., 1991).

Collagen matrices are obtained by lyophilisation (freeze-drying) of collagen solution/gel. The specificity of porous structure is the very low specific density, of approximately 0.02-0.3 g/cm^3 (Albu 2011, Zilberman & Elsner, 2008; Stojadinovic et al., 2008; Trandafir et al., 2007). The matrix porous structure depends significantly on collagen concentration, freezing rate, size of gel fibrils and the presence or absence of cross-linking agent (Albu et al., 2010b). The collagen matrix morphological structure is important, influencing the hydrophilicity, drug diffusion through network, degradation properties and interaction with cells. Figure 3 shows characteristic pore structure with a large variation in average pore diameter in collagen matrices.

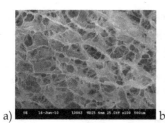 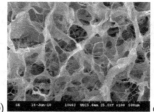 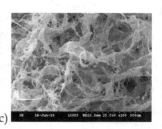

Fig. 3. SEM images of collagen matrices: (a) freeze at -40°C, (b) freeze at -10°C and (c) cross-linked with 0.25% glutaraldehyde.

Although the matrices presented in Fig. 3a,b have the same composition, their structure is different. Therefore, the low temperature (e.g. -40°C) induces about 10 times smaller pore sizes than higher temperature (e.g. -10°C). It can be noticed that lower freezing temperature produces more homogeneous samples than those obtained at high freezing temperature. Major differences of pore size and shape appear between un-cross-linked and cross-linked samples, the most homogeneous matrix with the smallest and inner pores being the un-cross-linked obtained at lowest freezing temperature.

Hydrophilic properties expressed by absorbing water and its vapor, are characteristic for collagen matrices, which can absorb at least 1500% water. Permeability for ions and macromolecules is of particular importance for tissues which are not based only on the vascular transport of nutrients. Diffusion of nutrients into the interstitial space ensures survival of the cells, continued ability to grow and to synthesize extracellular matrix specifically for tissue.

The infrared spectra of collagen exhibit several features characteristic for the molecular organization of its molecules: amino acids linked together by peptide bonds give rise to infrared active vibration modes amide A and B (about 3330 and 3080 cm-1, respectively) and amide I, II, and III (about 1629-1658 cm-1, 1550-1560 cm-1 and 1235-1240 cm-1, respectively) (Sionkowska et al., 2004).

Hydrothermal stability of collagen is characterized by its contraction when heated in water at a certain temperature at which the conformational transition of molecules from the triple helix statistic coil take place (Li, 2003).

Thermal behavior of collagen matrices depends on the number of intermolecular bonds. Generally, the number of bonds is higher, the shrinkage temperature is higher and the biomaterial is more stable *in vivo*.

Another method commonly used to assess the *in vivo* stability of collagen biomaterials, is the *in vitro* digestion of matrix with collagenase and other proteinases (trypsin, pepsin) (Li, 2003). Biodegradability of collagen matrices is dependent on the degree of cross-linking.

Collagen can form a variety of homogeneous collagen composites with ceramics, drugs, natural or synthetic polymers. The obtaining methods involve chemical cross-linking, physical loading and co-precipitation followed by free-drying, freeze-drying or electrospinning.

The most recent collagen composites used as medical devices, artificial implants, supports for drug release and scaffolds for tissue regeneration are presented in Table 1.

Collagen composites containing physiologically active substances acting as drug delivery systems (DDS) are discussed in Section 3.

Type of composite	Type of component from composite	Composite form
Collagen-natural polymer	Hyaluronic acid (Davidenko et al., 2010)	Matrix, membrane, hydrogel, fibers
	Silk fibroin (Zhou et al., 2010)	Membrane, fibers, matrix, microtubes
	Chondroitin-6-suphate (Stadlinger et al., 2008)	Tube, matrix
	Elastin (Skopinska-Wisniewska et al., 2009)	Tube, film, fibers, matrix
	Alginate (Sang et al., 2011)	Spongious, filler for bone,
	Chitosan (Sionkowska et al., 2004)	Matrix, membrane, tubular graft, nanofibers, hydrogel,
	Heparin (Stamov et al., 2008)	Matrix
Collagen-synthetic polymer	Poly-L-lactide (PLLA) (Chen et al., 2006)	Coating for composite
	Poly-lactic-co-glycolic-acid (PLGA) (Wen et al., 2007)	Fibers, matrix, coated tube
	ε-caprolactone (Schnell et al., 2007)	Nanofibers
	Poly(ethylene-glycol) (PEG) (Sionkowska et al., 2009)	Films, fibers
Collagen-ceramic	Calcium phosphates (Hong et al., 2011)	Matrix, filler
	Hydroxyapatite (Zhang et al., 2010; Hoppe et al., 2011)	Matrix, filler
	Tricalciumphosphate (Gotterbarm et al., 2006)	Matrix, filler

Table 1. Collagen-based composites

3. Collagen-based drug delivery systems

Nowdays, the field of drug delivery from topical biopolymeric supports has an increased development due to its advantages compared to the systemic administration. These biopolymers can release adequate quantities of drugs, their degradation properties being adjustable for a specific application that will influence cellular growth, tissue regeneration, drug delivery and a good patient compliance (Zilberman & Elsner, 2008). Among the biopolymers, collagen is one of the most used, being a suitable biodegradable polymeric support for drug delivery systems, offering the advantage of a natural biomaterial with haemostatic and wound healing properties (Lee et al., 2001).

Studies with collagen as support showed that *in vivo* absorption and degradability on the one hand and drug delivery on the other hand are controlled by the collagen chemical or physical cross-linking performed in order to control the delivery effect (Albu, 2011).

Among the incorporated drugs in the collagen biomaterials various structures are mentioned: antibiotics and antiseptic (tetracycline, doxicicline, rolitetracycline, minocycline, metronidazole, ceftazidine, cefotaxime, gentamicin, amikacin, tobramycin, vancomycin, clorhexidine), statines (rosuvastatin), vitamines (riboflavine), parasympathomimetic alkaloid (pilocarpine) etc. (Zilberman & Elsner, 2008; Goissis & De Sousa, 2009; Yarboro et al., 2007).

The most known collagen-based drug delivery systems are the hydrogels and matrices.
The literature in the field reveals the importance of modeling the drug release kinetics from systems with topical application. The topical preparations with antibiotics, anti-inflammtories, antihistaminics, antiseptics, antimicotics, local anaesthetics must have a rapid realease of the drug. The release kinetics has to balance the advantage of reaching a therapeutical concentration with the disadvantage of toxic concentrations accumulation (Ghica, 2010).

As far as the drug delivery kinetics from semisolid/solid systems generally is concerned, it has been widely studied only in the case of the hydrogels having quasi-solid structure (Lin & Metters, 2006; Albu et al., 2009b).

In the case of the matrices, there is scarce literature on the delivery and the delivery mechanism of the drug from such systems. In Fig. 4 the drug delivery from a spongious collagen support is schematically presented.

The delivery of the drug from polymeric formulations is controlled by one or more physical processes including: polymer hydration through fluid, swelling to form a gel, drug diffusion through the gel formed and eventual erosion of the polymeric gel. It is possible that, for the sponges, the swelling, erosion and the subsequent diffusion kinetics play an important role in the release of the drug from these systems upon contact with biological fluids (cutaneous wound exsudate/gingival crevicular fluid). Upon contact of a dry sponge with the wet surface at the application site, biological fluid from that region penetrates the polymer matrix. Thus, the solvent molecules' internal flux causes the subsequent sponge hydration and swelling and the formation of a gel at the application site surface. The swelling noticed is due to the polymeric chains solvation that leads to an increase of the distance between the individual molecules of the polymer (Peppas et al., 2000; Boateng et al., 2008).

For some of the spongious forms the drug release mechanism has been explained through the hydrolytic activity of the enzymes existing in biological fluids, different mathematical models of the collagen sponges' enzymatic degradation being suggested (Metzmacher et al., 2007; Radu et al., 2009).

It was shown that in an aqueous medium the polymer suffers a relaxation process having as result the direct, slow erodation of the hydrated polymer. It is possible that its swelling and dissolution happen at the same time as in the sponges' situation, each of these processes contributing to the global release mechanism. However, the quantity of the drug released is generally determined by the diffusion rate of the medium represented by biological fluid in the polymeric sponge. Factors such as polymeric sponge erosion after water diffusion and the swelling in other dosage forms are the main reason of kinetics deviation square root of time (Higuchi type, generally specific to the hydrogels as such) (Boateng et al., 2008).

Different methods have been suggested for the investigation of the drug controlled release mechanisms that combine the diffusion, the swelling and the erosion. It is assumed that the collagen sponge is made of a homogeneous polymeric support where the drug (dissolved or suspended) is present in two forms: free or linked to the polymeric chains. The drug as free form is available for diffusion, through the desorption phenomenon, for immediate release in a first stage, this being favored by the sponge properties behaving as partially open porosity systems. The drug amount partially imobilized in collagen fibrillar structure will be gradually released after the diffusion of the biologic fluid inside the sponge, followed by its swelling and erosion on the basis of polymers reaction in solution theory. This sustained release is favored by the matrix properties to act as partially closed porosity systems, as well as by the collagen sponge tridimensional structure, which is a barrier between the drug in the sponge and the release medium (Singh et al., 1995; Friess, 1998; Wallace & Rosenblatt, 2003; Ruszczak & Friess; 2003).

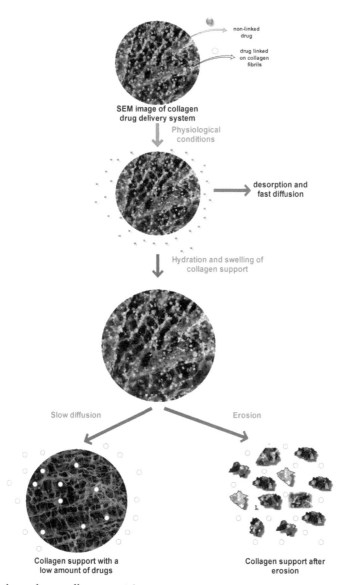

Fig. 4. Drug release from collagen matrix

In addition, the drug release kinetics can be influenced by the different chemical treatments that affect the degradation rate or by modifications of sponge properties (porosity, density) (M. Grassi & G. Grassi, 2005).

Among the chemical methods we can mention the cross-linking techniques. Thus, the different *in vivo* and *in vitro* behaviour, including the drug delivery profiles, can be obtained if the product based on collagen suffer in addition cross-linking with different cross-linking agents. Among these agents, the most known and used is the glutaraldehyde that forms a link between the ε-amino groups of two lateral lysin chains. It was demonstrated that the treatment with glutaraldehyde reduces the collagen material immunogenicity, leading at the same time to the increase of resistance to enzymatic degradation (Figueiro et al., 2006).

Concerning the preparation of sponges with different porosities, those can be obtained by modifying the temperature during the collagen sponges lyophilization process (Albu et al., 2010a).

To understand the release process, both from hydrogels and from collagen sponges, and to establish the drug release mechanism implicitly, a range of kinetic models is used (Peppas, Higuchi, zero order). The general form of the kinetic equation through which the experimental kinetic data are fitted is the following: (eq. 1)

$$\frac{m_t}{m_\infty} = k \cdot t^n \tag{1}$$

where m_t is the amount of drug released at time t, m_∞ is the total drug contents in the designed collagen hydrogels, m_t/m_∞ is the fractional release of the drug at the time t, k is the kinetic constant, reflecting the structural and geometrical properties of the polymeric system and the drug, and n is the release exponent, indicating the mechanism of drug release.

If n=0.5 the release is governed by Fickian diffusion (the drug diffusion rate is much lower than the polymer relaxation rate, the amount of drug released being proportional to the release time square root, corresponding to Higuchi model). If n=1 the release is controlled by surface erosion (the drug diffusion rate is much higher than the polymer relaxation rate, the amount of drug released being proportional with the release time, corresponding to zero order model). If 0.5<n<1, the drug release mechanism is of non-Fickian type diffusion, the drug diffusion rate and the polymer relaxation rate being roughly equal. In this case the release is not based only on diffusion, being also associated with other release mechanisms due to the complex processes previousely described (Teles et al., 2010; Higuchi, 1962; Peppas et al., 2000; Singh et al., 1995; Ho et al., 2001).

The studied literature shows that Peppas, Higuchi and zero order models do not explain the mechanisms involved in the kinetic processes in the case of sponge forms, because the value of the apparent release order value (n) does not fit between the limits imposed by these aforesaid models, having values much lower than 0.5. This is why an extension of Peppas model to the Power law model (0<n<1) is generally applied in order to elucidate the complex kinetic mechanisms involved in the drug release from such natural supports. Practically, n value includes characteristics of each particular model previously described (Ghica et al., 2009; Albu et al., 2009a; Albu et al., 2010b; Phaechamud & Charoenteeraboon, 2008; Natu et al., 2007).

Our studies showed values of n equal to 0.5 for different collagen-based hydrogels with doxyciclyne, uncross-linked or cross-linked with glutaraldehyde, which confirms the respect of Higuchi model concerning the drug release from semisolid supports (Albu et al., 2009b). On the contrary, n values for doxycycline release from spongious supports were inferior to 0.5 (Ghica et al., 2009; Albu et al., 2010a; Albu et al., 2010b).

4. Collagen-based scaffolds for tissue engineering

It is known that collagen is the major component of the extracellular matrix of most tissues. As a natural molecule, collagen possesses a major advantage in being biodegradable, biocompatible presented low antigenicity, easily available and highly versatile. Collagen provides structural and mechanical support to tissues and organs (Gelse et al., 2003) and fulfils biomechanical functions in bone, cartilage, skin, tendon, and ligament. For this reason, allogenic and xenogenic collagens have been long recognized as one of the most useful biomaterials. Collagen can be prepared in a number of different forms with different application: shields used in ophthalmology (Rubinstein, 2003; Yoel & Guy, 2008) matrices for burns/wounds (Keck et al., 2009; Wollina et al., 2011), gel formulation in combination with liposomes for sustained drug delivery (Wallace & Rosenblatt, 2003; Weiner et al., 1985; Rao, 1996), as controlling material for transdermal delivery (Rao, 1996; Thacharodi & Rao, 1996), nanoparticles for gene delivery (Minakuchi et al., 2004) and basic matrices for cell culture systems. Therefore thin sheets and gels are substrates for smooth muscle (Dennis et al., 2007; Engler et al., 2004), hepatic (Hansen & Albrecht, 1999; Ranucci et al., 2000), endothelial (Albu et al., 2011; Deroanne et al., 2001; Titorencu et al., 2010), and epithelial cells (Haga et al., 2005), while matrices are often used to engineer skeletal tissues such as cartilage (Stark et al., 2006; Schulz et al., 2008), tendon (Gonçalves-Neto et al., 2002; Kjaer, 2004) and bone (Guille et al., 2005).

It is known that the goal of tissue engineering (TE) is to repair and restore damaged tissue function. The three fundamental "tools" for both morphogenesis and tissue engineering are responding cells, scaffolds and growth factors (GFs - regulatory biomolecules, morphogens), which, however, are not always simultaneously used (Badylak & Nerem, 2010; Berthiaume et al., 2011) (Fig. 5).

In tissue engineering, matrices are developed to support cells, promoting their differentiation and proliferation in order to form a new tissue. Another important aspect for the generation of 3D cell matrix constructs suitable for tissue regeneration is represented by cell seeding. Besides the seeding technique, the cellular density is a crucial factor to achieve a uniform distribution of a number of cells which is optimal for new tissue formation (Lode et al., 2008). Such strategies allow producing of biohybrid constructs that can be implanted in patients to induce the regeneration of tissues or replace failing or malfunctioning organs.

The advantage of tissue engineering is that small biopsy specimens can be obtained from the patient and cells can be isolated, cultured into a structure similar to tissue or organs in the living body, expanded into large numbers (Bruder & Fox, 1999; Levenberg & Langer, 2004; Mooney & Mikos, 1999; Service, 2005) and then transplanted into the patients.

The recent advances in collagen scaffold biomaterials are presented as follows:

Wound dressing and delivery systems

In the treatment of wounds the collagen-based dressings are intensely used. There are many studies which attest the benefits of topical collagen matrices on the wound healing (Inger & Richard, 1999; Ruszczak, 2003; Shih-Chi et al., 2008).

It is known that collagen matrices absorb excess wound exudate or sterile saline, forming a biodegradable gel or sheet over the wound bed that keeps the balance of wound moisture environment, thus promoting healing (Hess, 2005). Also, collagen breakdown products are chemotactic for a variety of cell types required for the formation of granulation tissue. Nowadays, many types of skin substitutes using living cells have been used clinically (Table 2).

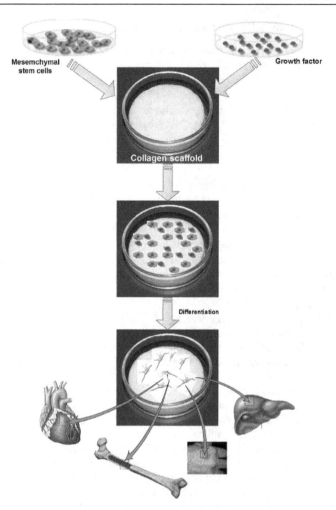

Fig. 5. Strategies for tissue engineering

Classification	Tissue replaced	Layers
TranCyte®	Epidermal	Silicone Nylon mesh Collagen seeded with neonatal fibroblasts
PermaDerm™	Dermal	Autologous fibroblasts in bovine collagen matrix with autologous keratinocytes
Apligraft®	Epidermal and dermal	Neonatal keratinocytes Collagen seeded with neonatal fibroblasts
OrCell™	Epidermal and dermal	Collagen (bovine type I) seeded with allogenic fibroblasts and keratinocytes

Table 2. Classification of collagen substitutes with living cells

These biomaterials can be divided into three groups depending on the type of layer cells which are substituted. The first type consists of grafts of cultured epidermal cells with no dermal components. The second type has only dermal components. The third type is a bilayer containing both dermal and epidermal elements.

Bone defects

Bone development and regeneration occurs as a result of coordinated cell proliferation, differentiation, migration, and remodeling of the extracellular matrix. In bone tissue engineering collagen scaffolds play an essential role in supporting bone regeneration. The implantation of these 3D biomaterials is necessary when osteochondral defects reach an important volume or when autografts have to be avoided for practical or pathological reasons. In order to promote bone healing scaffolds must have some properties: to promote the differentiation of immature progenitor cells into osteoblasts (*osteoinduction*), to induce the ingrowth of surrounding bone (*osteoconduction*) and to be integrated into the surrounding tissue (*osseointegration*) (Dickson et al., 2007). However there is still some ambiguity regarding the optimal porosity and pore size for a 3D bone scaffold. A literature review indicates that a pore size in the range of 10–400 μm may provide enough nutrient and osteoblast cellular infusion, while maintaining structural integrity (Bignon et al., 2003; Holmbom et al., 2005; Woodard et al., 2007).

Collagen scaffolds have the advantage of facilitating cell attachment and maintaining differentiation of cells (Fig. 6). Resorbable collagen sponges have been successfully used as carriers of BMP-2, BMP-4 and BMP-7 but they have the disadvantages of a fast degradation rate and low mechanical strength (Bessa et al., 2008; Higuchi et al., 1999; Huang et al., 2005; Huang et al., 2005; Kinoshita et al., 1997;).

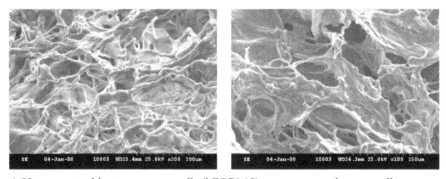

Fig. 6. Human osteoblasts precursor cells (hFOB1.19) grown seven days on collagen scaffolds

In order to increase mechanical strength and to improve the release of growth factors a combination between collagen and other natural polymer has been used such as chitosan (Arpornmaeklong et al., 2007), dextran (Fig. 7) or glycosaminoglycans (Harley & Gibson, 2008; Wang et al., 2010).

Another combination of collagen scaffolds is represented by mineralization with calcium phosphate (Du et al., 2000; Harley et al., 2010) and/or on cross-linking with other substances like hydroxyapatite (Dubey & Tomar, 2009; Liao et al., 2009.) or bushite (Tebb et al., 2006).

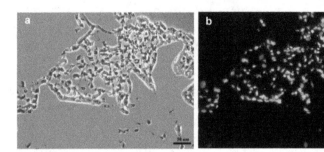

Fig. 7. Osteosarcoma cells (MG 63) grown on collagen-dextran scaffold: a – phase contrast, b – Hoechst nuclear staining

Urogenital system

Injuries of the genitourinary system can lead to bladder damage. Treatment in most of these situations requires eventual reconstructive procedures that can be performed with native non-urologic tissues (skin, gastrointestinal segments or mucosa), heterologous tissues or substances (bovine collagen) or artificial materials (Atala, 2011). Acellular collagen scaffolds were used in the treatment of bladder augmentation (Akbal et al., 2006; Liu et al., 2009; Parshotamet al., 2010) and urethral stricture (el-Kassaby et al., 2008; Farahat et al., 2009). Also collagen-composite scaffolds populated with the patient's own urothelial and muscle cells or self-assembled fibroblast sheets represent a promising strategy for bladder augmentation (Bouhout et al., 2010; Magnan et al., 2006). Trials of urethral tissue replacement with processed collagen matrices are in progress, and bladder replacement using tissue engineering techniques are intensely being studied (Atala et al., 2006).

Scaffolds for hepatic cells

Recent new strategies for treating liver diseases, including the extracorporeal bioartificial liver device and hepatocyte transplantation represent the future in hepatic diseases treatment. Recent advances in the field of tissue engineering have demonstrated that type I collagen matrices induced the differentiation of hepatic stem-like cells into liver epithelial cells and that biodegradable collagen matrices provide an appropriate microenvironment for hepatocytic repopulation (Uneo et al., 2004;Ueno et al., 2005.); also, a combination between collagen-chitosan-heparin scaffolds was used in order to obtain bioartificial liver (Xing et al., 2005).

Cornea and neural cells

Bioengineered corneas are substitutes for human donor tissue that are designed to replace part or the full thickness of damaged or diseased corneas. Collagen has been used successfully in reconstruction of artificial cornea alone by delivery of limbal epithelial stem cells to damaged cornea (Builles et al., 2010; De Miguel et al., 2010) or in combination with glycosaminoglycan (GAG) molecules (Auxenfans et al., 2009). The combination of collagen biomaterials and stem cells could be a valuable strategy to treat corneal defects also. Other strategies in collagen-based corneal scaffolds include the utilization of recombinant human collagen (Dravida et al., 2008; Griffith et al., 2009), the secretion of collagen by the fibroblasts themselves (Carrier et al., 2008) and surface modification to reduce extensive endothelialization (Rafat et al., 2009).

Nerve repairing

One of the major challenges in neurology is to be able to repair severe nerve trauma. It was observed that collagen scaffolds is a suitable nerve guidance material (Han et al., 2010) Most collagen nerve guides are engineered from crosslinked collagen with tubular shape such as commercially available NeuraGen® from Integra™. Recent tissue engineering strategies involve addition of neurotrophic factors into collagen scaffolds (Sun et al., 2007; Sun et al., 2009) and cell delivery (Bozkurt et al., 2009; Kemp et al., 2009) in order to attempt to enhance nerve guides.

5. Conclusion

Collagen biomaterials as matrices, hydrogels, composites have already been proved to be effective in tissue repairing, in guiding functional angiogenesis and in controlling stem cell differentiation. Also, collagen-based drug delivery systems were studied and their mechanisms of release were determined. Based on such good results, the promising next generation of engineered tissues is relying on producing scaffolds which can prolong the release rate of growth factors or cells in order to increase their therapeutic effect. This justifies the importance of drug delivery in tissue engineering applications.

6. Acknowledgment

The authors would like to thank Dr. Viorica Trandafir for guiding their steps in the collagen biomaterial field and for precious information from her own experience.

7. References

Akbal, C.; Lee, S.D.; Packer, S.C.; Davis, M.M.; Rink, R.C. & Kaefer, M. (2006). Bladder augmentation with acellular dermal biomatrix in a diseased animal model. *Journal of Urology*, Vol.176, No.4, (October 2006), pp. 1706–1711, ISSN 00225347

Albu, M.G. & Leca, M. (2005). Rheological behaviour of some cross-linked collagen hydrogels for drug delivery use. *European Cells and Materials*, Vol.10, Suppl.1, pp. 8, ISSN 1473-2262

Albu, M.G.; Ghica, M.V.; Giurginca, M.; Trandafir, V.; Popa, L. & Cotrut, C. (2009a). Spectral characteristics and antioxidant properties of tannic acid immobilized on collagen drug-delivery systems. *Revista de Chimie – Bucharest*, Vol.60, No.7, (July 2009), pp. 666-672, ISSN 0034-7752

Albu, M.G.; Ghica, M.V.; Popa, L.; Leca, M. & Trandafir, V. (2009b). Kinetics of in vitro release of doxycycline hyclate from collagen hydrogels, *Revue Roumaine de Chimie*, Vol.54, No.5, (May 2009), pp. 373-379, ISSN 0035-3930

Albu, M.G.; Ghica, M.V.; Ficai, A.; Titorencu, I..& Popa, L. (2010a). The influence of freeze-drying process on porosity and kinetics release of collagen-doxycycline matrices. *Proceedings of 3rd International Conference on Advanced materials and systems ICAMS*, ISBN 2068-0783, Bucharest, Romania, September, 2010

Albu, M.G.; Ficai, A. & Lungu, A. (2010b). Preparation and characterization of collagen matrices obtained at different freezing temperatures, *The Leather and Footwear Journal*, Vol. 10, No. 3, (September 2010), pp.39-50, ISSN 1583-4433

Albu, M.G.; Ghica, M.V.; Leca, M.; Popa, L.; Borlescu, C.; Cremenescu, E.; Giurginca, M. & Trandafir, V. (2010c). Doxycycline delivery from collagen matrices crosslinked with tannic acid. *Molecular Crystals & Liquid Crystals*, Vol.523, pp. 97/[669]-105[677], ISSN 1542-1406

Albu, M.G. (2011). *Collagen gels and matrices for biomedical applications*, LAP LAMBERT Academic Publishing GmbH & Co. KG, ISBN 978-3-8443-3057-1, Saarbrücken, Germany

Albu, M.G.; Titorencu, I. & Chelaru, C. (2011). The stability of some collagen hydrogels, *The Leather and Footwear Journal, Vol. 11, No.1, (March 2011), pp.11-20, ISSN* 1583-4433

Angele, P.; Abke, J.; Kujat, R.; Faltermeier, H.; Schumann, D.; Nerlich, M.; Kinner, B.; Englert, C.; Ruszczak, Z.; Mehrl, R. & Mueller, R. (2004), Influence of different collagen species on physico-chemical properties of crosslinked collagen matrices, *Biomaterials*, Vol.25, No.14, (June 2004), pp.2831-2841, ISSN 0142-9612

Arpornmaeklong, P.; Suwatwirote, N.; Pripatnanont, P. & Oungbho, K. (2008). Growth and differentiation of mouse osteoblasts on chitosan–collagen sponges. *Journal of Oral Maxillofacial Surgery*, Vol.36, No.4, (April 2007), pp. 328–337, ISSN 02782391

Atala A.; Bauer, S.B.; Soker, S.; Yoo, J.J. & Retik, A.B. (2006). Tissue-engineered autologous bladders for patients needing cystoplasty. *Lancet*, Vol. 367, No.9518, (April 2006), pp. 1241–1246, ISSN 0140-6736

Atala, A. (2011). Tissue engineering of human bladder. *British Medical Bulletin*, Vol.97, No.1, (February 2011), pp. 1–24, ISSN 1471-8391

Auxenfans, C.; Builles, N.; Andre, V.; Lequeux, C.; Fievet, A.; Rose, S.; Braye, F.M.; Fradette, J.; Janin-Manificat, H.; Nataf, S.; Burillon, C & Damour, O. (2009). Porous matrix and primary-cell culture: a shared concept for skin and cornea tissue engineering. *Pathologie Biologie (Paris)*, Vol.57, No.4, (June 2009), pp. 290-298, ISSN 0369-8114

Badylak, S.F. & Nerem R.M. (2010). Progress in tissue engineering and regenerative medicine. *PNAS, Vol.*107, No.8, (February 2010), pp. 3285–3286, ISSN: 1529-24

Bazin, S. & Delaumay, A. Preparation of acid and citrate soluble collagen, In: *The Methodology of Connective Tissue Research*, Hall D.A., pp.13-18, Joynson–Bruvvers, ISBN 0903848066, Oxford

Berthiaume, F.; Maguire, T.J. & Yarmush, M.L. (2011). Tissue Engineering and Regenerative Medicine: History, Progress, and Challenges. *Chemical and Biomolecular Engineering* Vol.2, (June 2011), pp. 1-18, ISSN: 1307-7449

Bessa, P.C.; Casal, M. & Reis, R.L. (2008). Bone morphogenetic proteins in tissue engineering: the road from laboratory to clinic, part II (BMP delivery). *Journal of Tissue Engineering and Regenerative Medicine*, Vol.2, No.2-3, (March - April 2008), pp. 81–96, ISSN 1932-6254

Bignon, A.; Chouteau, J.; Chevalier, J.; Fantozzi, G.; Carret, J.P.; Chavassieux, P.; Boivin, G.; Melin, M. & Hartmann, D. (2003). Effect of micro- and macroporosity of bone substitutes on their mechanical properties and cellular response. *Journal of Matererial Science Material Medicine*, Vol.14, No.12, (December 2003), pp.1089–1097, ISSN 0957-4530

Boateng, J.S.; Matthews, K.H.; Stevens, H.N.E. & Eccleston, G.M. (2008). Wound healing dressings and drug delivery systems: a rewiev. *Journal of Pharmaceutical Sciences*, Vol.97, No.8, (August 2008), pp. 2892-2923, ISSN 0022-3549

3ouhout, S.; Perron, E.; Gauvin, R.; Bernard, G.; Ouellet, G.; Cattan, V. & Bolduc, S. (2010). *In vitro* reconstruction of an autologous, watertight and resistant vesical equivalent. *Tissue Engineering Part A*, Vol.16, No.5, (May 2010), pp.1539-1548, ISSN 1937-3341

3ozkurt, A.; Deumens, R.; Beckmann, C.; Olde Damink, L.; Schugner, F.; Heschel, I.; Sellhaus, B.; Weis, J.; Jahnen-Dechent, W.; Brook, G.A. & Pallua, N. (2009). *In vitro* cell alignment obtained with a Schwann cell enriched microstructured nerve guide with longitudinal guidance channels. *Biomaterials*, (January 2009), Vol.30, No.2, pp.169–179, ISSN 0142-9612

3ruder, S.P. & Fox, B.S. (1999). Tissue engineering of bone. Cell based strategies. *Clinical Orthopopaedics*, Vol.367, (October 1999), pp: S68–S83, ISSN 0095-8654

3uilles, N.; Janin-Manificat, H.; Malbouyres, M.; Justin, V.; Rovère, M.R,; Pellegrini, G.; Torbet, J.; Hulmes, D.J.; Burillon, C.; Damour, O. & Ruggiero F. (2010). Use of magnetically oriented orthogonal collagen scaffolds for hemi-corneal reconstruction and regeneration. *Biomaterials*, Vol.31, No.32, (November 2010), pp. 8313-8322, ISSN 0142-9612

Carrier, P.; Deschambeault, A.; Talbot, M.; Giasson, C.J.; Auger, F.A.; Guerin, S.L. & Germain, L. (2008). Characterization of wound reepithelization using a new human tissue-engineered corneal wound healing model. *Invesigation. Ophthalmology & Visual Science*, Vol.49, No.4, (April 2008), pp. 1376–1385, ISSN 1552-5783

Chen, S.; Hirota, N.; Okuda, M.; Takeguchi, M.; Kobayashi, H.; Hanagata, N. & Ikoma, T. (2011). Microstructures and rheological properties of tilapia fish-scale collagen hydrogels with aligned fibrils fabricated under magnetic fields, *Acta Biomaterialia*, Vol.7, No.2, (February 2011), pp. 644-652, ISSN 1742-7061

Chen, Y.; Mak, A.F.T; Wang, M.; Li, J. & Wong, M.S. (2006). PLLA scaffolds with biomimetic apatite coating and biomimetic apatite/collagen composite coating to enhance osteoblast-like cells attachment and activity, *Surface and Coatings Technology*, Vol.201, No.3-4, (October 2006), pp.575-580, ISSN 0257-8972

Cioca, Gh. (July 1981). *Process for preparing macromolecular biologically active collagen*, Patent No. 4279812, US

Davidenko, N.; Campbell, J.J.; Thian, E.S.; Watson, C.J. & Cameron, R.E. (2010). Collagen–hyaluronic acid scaffolds for adipose tissue engineering, *Acta Biomaterialia*, Vol. 6, No.10, (October 2010), pp.3957-3968, ISSN 1742-7061

De Miguel, M.P.; Alio, J.L.; Arnalich-Montiel, F.; Fuentes-Julian, S.; de Benito-Llopis, L.; Amparo, F. & Bataille, L. (2010). Cornea and ocular surface treatment. *Current Stem Cell Research Therapy*, Vol.5, No.2, (June 2010), pp. 195-204, ISSN 1574-888X

Deroanne, C.F.; Lapiere, C.M. & Nusgens, B.V. (2001). In vitro tubulogenesis of endothelial cells by relaxation of the coupling extracellular matrix-cytoskeleton. *Cardiovascular Research*, Vol.49, No.3, (September 2001), pp. 647-658, ISSN 0008-6363

Dickson, G.; Buchanan, F.; Marsh, D.; Harkin-Jones, E.; Little, U. & McCaigue M. (2007). Orthopaedic tissue engineering and bone regeneration. *Technology and Health Care*, Vol.15, No.1, (January 2007), pp.57-67, ISSN 0928-7329

Doillon, C. J. (1992). Skin replacement using collagen extracted from bovine hide, *Clinical Materials*, Vol.9, No.3-4, (April 2004), pp.189-193, ISSN 0267-6605

Dravida, S.; Gaddipati, S.; Griffith, M.; Merrett, K.; Lakshmi Madhira, S. & Sangwan, V.S. (2008). A biomimetic scaffold for culturing limbal stem cells: a promising alternative for clinical transplantation. *Journal of Tissue Engineering Regenerative Medicine*, Vol.2, No.5, (July 2008), pp. 263–271, ISSN 1932-7005

Du, C.; Cui, F.Z.; Zhang, W.; Feng, Q.L.; Zhu, X.D. & de Groot, K. (2000). Formation of calcium phosphate/collagen composites through mineralization of collagen matrix. *Journal of Biomedical Materials Research Part A*, Vol.50, No.4, (April 2000), pp. 518–527, ISSN 1549-3296

Dubey, D.K. & Tomar, V. (2009). Role of the nanoscale interfacial arrangement in mechanical strength of tropocollagen-hydroxyapatite-based hard biomaterials. *Acta Biomaterialia*, Vol.5, No.7, (September 2009), pp. 2704–2716, ISSN 1742-7061

Edwards, G.A. & Roberts, G. (1992), Development of an ovine collagen-based composite biosynthetic vascular prosthesis, *Clinical Materials*, Vol.9, No.3-4, (April 2004), pp.211-223, ISSN 0267-6605

el-Kassaby, A.; AbouShwareb, T. & Atala, A. (2008). Randomized comparative study between buccal mucosal and acellular bladder matrix grafts in complex anterior urethral strictures. *Journal of Urology*, Vol.179, (April 2008), 1432–1436, ISSN 00225347

Engler, A.; Bacakova, L.; Newman, C.; Hategan, A.; Griffin, M. & Discher, D. (2004). Substrate Compliance versus Ligand Density in Cell on Gel Responses. *Biophysical Journal*; Vol.86, No.1, (January 2004), pp. 617–628, ISSN 0006-3495

Farahat, Y.A.; Elbahnasy, A.M.; El-Gamal, O.M.; Ramadan, A.R.; El-Abd, S.A. & Taha, M.R. (2009). Endoscopic urethroplasty using small intestinal submucosal patch in cases of recurrent urethral stricture: A preliminary study. *Journal of Endourology*, Vol.23, No.12, (December 2009), pp. 2001–2005, ISSN 0892-7790

Fielding, A.M. (1976). Preparation of neutral salt soluble collagen, In: *The Methodology of Connective Tissue Research*, Hall D.A., pp.9-12, Joynson–Bruvvers, ISBN 0903848066, Oxford

Figueiro, S.D.; Macedo, A.A.M.; Melo, M.R.S.; Freitas, A.L.P.; Moreira, R.A.; De Oliveira, R.S.; Goes, J.C. & Sombra, A.S.B. (2006). On the dielectric behaviour of collagen-algal sulfated polysaccharide blends: effect of glutaraldehyde crosslinking. *Biophysical Chemistry*, Vol.120, No.2, (March 2006), pp. 154-159, ISSN 0301-4622

Friess, W. (1998). Collagen – biomaterial for drug delivery. *European Journal of Pharmaceutics and Biopharmaceutics*, Vol.45, No.2, (March 1998), pp.113-136, ISSN 0939-6411

Ghica, M.V. (2010). *Physico-chemical and biopharmaceutical elements of semisolid systems with topical action. Applications to indomethacin hydrogels*, ISBN 978-606-521-436-1, Printech, Bucharest, Romania

Ghica, M.V.; Albu, M.G.; Popa, L.; Leca, M.; Brazdaru, L.; Cotrut, C. & Trandafir, V. (2009). Drug delivery systems based on collagen-tannic acid matrices. *Revue Roumaine de Chimie*, Vol.54, No.11-12, (November-December 2009), pp. 1103-1110, ISSN 0035-3930

Goissis, G. & De Sousa, M.H. (2009). Characterization and in vitro release studies of tetracycline and rolitetracycline imobilized on anionic collagen membranes. *Materials Research*, Vol.12, No.1, pp. 69-74, ISSN 1516-1439

Gonçalves-Neto, J.; Witzel, S.S.; Teodoro, W.R.; Carvalho-Júnior, A.E.; Fernandes, T.D. & Yoshinari, H.H. (2002). Changes in collagen matrix composition in human posterior tibial tendon dysfunction. *Joint Bone Spine*, Vol.69, No.2, (Martch 2002), pp.189-194, ISSN: 1297-319X

Gotterbarm, T; Richter, W.; Jung, M.; Vilei, S.B.; Mainil-Varlet, P.; Yamashita, T & Breusch, S.J. (2006). An in vivo study of a growth-factor enhanced, cell free, two-layered collagen–tricalcium phosphate in deep osteochondral defects, *Biomaterials*, Vol. 27, No.18, (June 2006), pp.3387-3395, ISSN 0142-9612

Grassi, M. & Grassi, G. (2005). Mathematical modelling and controlled drug delivery: matrix systems. *Current Drug Delivery*, Vol.2, No.1, (January 2005), pp. 97-116, ISSN 1567-2018

Griffith, M.; Jackson, W.B.; Lagali, N.; Merrett, K.; Li, F. & Fagerholm, P. (2009). Artificial corneas: A regenerative medicine approach. *Eye*, Vol.23, No.10, (October 2009), pp.1985-1989, 0950-222X

Guille, M.G.; Mosser, G.; Helary C. & Eglin, D. (2005). Bone matrix like assemblies of collagen: From liquid crystals to gels and biomimetic materials. *Micron*, Vol.36, No.7-8, (October-December 2005), pp. 602-608, ISSN 1463–9076

Haga, H.; Irahara, C.; Kobayashi, R.; Nakagaki, T. & Kawabata, K. (2005). Collective Movement of Epithelial Cells on a Collagen Gel Substrate. *Biophysical Journal*, Vol.88, No.3, (March 2005), pp. 2250–2256, ISSN 0006-3495

Han, Q.; Jin, W.; Xiao, Z.; Ni, H.; Wang, J.; Kong, J.; Wu, J.; Liang, W.; Chen, L.; Zhao, Y.; Chen, B. & Dai, J. (2010). The promotion of neural regeneration in an extreme rat spinal cord injury model using a collagen scaffold containing a collagen binding neuroprotective protein and an EGFR neutralizing antibody. *Biomaterials*, Vol.31, No.35, (December 2010), pp. 9212-9220, ISSN 0142-9612

Hansen, L.K. & Albrecht, J.H. (1999). Regulation of the hepatocyte cell cycle by type I collagen matrix: role of cyclin D1. *Journal of Cell Science*, Vol.112, (August 1999), pp. 2971-2981, ISSN 0021-9533

Harley B.A.C. & Gibson, L.J. (2008). *In vivo* and *in vitro* applications of collagen-GAG scaffolds. *Chemical Engineering Journal*, Vol.137, No.1, (March 2008), pp.102-121, ISSN 1385-8947

Harley, B.A.; Lynn, A.K.; Wissner-Gross, Z.; Bonfield, W.; Yannas, I.V. & Gibson, L.J. (2010). Design of a multiphase osteochondral scaffold. II. Fabrication of a mineralized collagenglycosaminoglycan scaffold. *Journal of Biomedical Materials Research Part A*, Vol. 92A, No.3, (March 2010), pp. 1066–1077, ISSN 1549-3296

Healy, K.E.; Rezania, A. & Stile, R.A. (1999). Designing biomaterials to direct biological responses, *Annals of the New York Academy of Sciences*, Vol. 875, (June 1999), pp.24-35, ISSN 1749-6632

Hess, C.T. (2005). *The art of skin and wound care documentation. Home Healthcare Nurse*, Vol.23, No.8, (August 2005), pp. 502-512, ISSN 0884741X

Higuchi, T.; Kinoshita, A.; Takahashi, K.; Oda, S. & Ishikawa, I. (1999). Bone regeneration by recombinant human bone morphogenetic protein-2 in rat mandibular defects. An experimental model of defect filling. *Journal of Periodontology*, Vol.70, No.9, (September 1999), pp. 1026–1031, ISSN 0022-3492

Higuchi, W.I. (1962). Analysis of data on the medicament release from ointments. *Journal of Pharmaceutical Sciences*, Vol.51, No.8, (August 1962), pp 802-804, ISSN 0022-3549

Ho, H.O.; Lin, C.W. & Sheu, M.T. (2001). Diffusion characteristics of collagen film. *Journal of Controlled Release*, Vol.77, No.1-2, (November 2001), pp. 97-105, ISSN 0168-3659

Holmbom, J.; Sodergard, A.; Ekholm, E; Martson, M.; Kuusilehto, A.; Saukko, P. & Penttinen, R. (2005). Long-term evaluation of porous poly(epsilon-caprolactone-co-

L-lactide) as a bone-filling material. *Journal of Biomedical Material Research A*, Vol.75, No.2, (November 2005), pp.308–315, ISSN 1549-3296

Hong, Y.J.; Chun, J-S. & Lee, W-K. (2011). Association of collagen with calcium phosphate promoted osteogenic responses of osteoblast-like MG63 cells, *Colloids and Surfaces B: Biointerfaces*, Vol.83, No.2, (April 2011), pp.245-253, ISSN 0927-7765

Hoppe, A.; Güldal, N.S. & Boccaccini, R.A. (2011). A review of the biological response to ionic dissolution products from bioactive glasses and glass-ceramics, *Biomaterials*, Vol.32, No.11, (April 2011), pp.2757-2774, ISSN 0142-9612

Huang, Y.C.; Kaigler, D.; Rice, K.G.; Krebsbach, P.H. & Mooney, D.J. (2005). Combined Angiogenic and Osteogenic Factor Delivery Enhances Bone Marrow Stromal Cell-Driven Bone Regeneration. *Journal of Bone and Mineral Research*, Vol.20, No.5, (May 2005), pp. 848–857, ISSN 1523-4681

Huang, Y.C.; Simmons, C.; Kaigler, D.; Rice, K.G. & Mooney, D.J. (2005). Bone regeneration in a rat cranial defect with delivery of PEI-condensed plasmid DNA encoding for bone morphogenetic protein-4 (BMP-4). *Gene Therapy*, Vol.12, No.5, (March 2005), pp. 418–426, ISSN 0969-7128

Hubbel, J.A. (1999). Bioactive biomaterials, *Current Opinion in Biotechnology*, Vol.10, No.2, (April 1999), pp.123-129, ISSN 0958-1669

Inger, A.J.S &. Clark, R.A.F. (1999). Cutaneus wound healing. *The New England Journal of Medicine* Vol.341, No.10, (September, 1999), pp.738-746, ISSN 1533-4406

Keck, M.; Haluza, D.; Burjak S.; Eisenbock, B.; Kamolz, L.-P. & Frey, M. (2009). Cultivation of keratinocytes and preadipocytes on a collagen-elastin scaffold (Matriderm®): First results of an in vitro study - Elastin Matrix (Matriderm®): Erste Ergebnisse einer In vitro Studie. *European Surgery*, Vol.41, No.4, (July 2009), pp.189-193, ISSN 1682-4016

Kemp, S.W.; Syed, S.; Walsh, W.; Zochodne, D.W. & Midha, R. (2009). Collagen nerve conduits promote enhanced axonal regeneration, schwann cell association, and neovascularization compared to silicone conduits. *Tissue Engineering Part A*, Vol.15, No.8, (August 2009), pp.1975–1988, ISSN 2152-4947

Ket, G.; Poschl, E. & Aigner, T. (2003). Collagens-structure, function, and biosynthesis. *Advanced Drug Delivery Review*, Vol.55, No.12, (November 2003), pp.1531–1546, ISSN 0169-409X

Kinoshita, A.; Oda, S.; Takahashi, K.; Yokota, S. & Ishikawa, I. (1997). Periodontal regeneration by application of recombinant human bone morphogenetic protein- 2 to horizontal circumferential defects created by experimental periodontitis in beagle dogs. *Journal of Periodontology*, Vol. 68, No.2, (February 1997), pp. 103–109, ISSN 0022-3492

Kjaer, M. (2004). Role of Extracellular Matrix in Adaptation of Tendon and Skeletal Muscle to Mechanical Loading. Physiology Review, Vol.84, No.2, (April 2004), pp. 649-698, ISSN 0031-9333

Kumar, M.S.; Kirubanandan. S.; Sripriya, R. & Sehgal, P.K. (2010). Triphala incorporated collagen sponge – a smart biomaterial for infected dermal wound healing. *Journal of Surgical Research*, Vol.158, No.1, (January 2010), pp. 162-170, ISSN 0022-4804

Lee, C.H.; Singla, A.& Lee, Y. (2001). Biomedical applications of collagen. *International Journal of Pharmaceutics*, Vol.221, No.1-2, (June 2001), pp. 1-22, ISSN 0378-5173

Levenberg, S. & Langer, R. (2004). Advances in tissue engineering. *Current Topics in Developmental Biology*, Vol.61, pp. 113–134, *ISBN* 9780080494340

Li, S.T.; Archibald, S.J.; Krarup, C. & Madison, R.D. (1991). The development of collagen nerve guiding conduits that promote peripheral nerve regeneration, In: *Biotechnology and Polymers*, Gebelein, C.G., pp.282-293, Plenum Press, ISBN 0-306-44049-0, New York

Li, S-T. (2003). Biologic Biomaterials: Tissue-Derived Biomaterials (Collagen), In: *Biomaterials Principle and Applications*, Park, J.B & Bronzino, J.D., pp.117-139, CRC Press, ISBN 0-8493-1491-7, Boca Raton, Florida

Liao, S.; Ngiam, M.; Chan, C.K. & Ramakrishna, S. (2009). Fabrication of nanohydroxyapatite/ collagen/osteonectin composites for bone graft applications. *Biomedical Materials,* Vol.4, No.2, (April 2009), pp. 025019, ISSN 1748-605X

Limpisophon, K.; Tanaka, M.; Weng, W.Y; Abe, S. & Osako, K. (2009), Characterization of gelatin films prepared from under-utilized blue shark (*Prionace glauca*) skin, *Food Hydrocolloids*, Vol. 23, No.7, (October 2009), pp.1993-2000, ISSN 0268-005X

Lin, C.C. & Metters, A.T. (2006). Hydrogels in controlled release formulations: Network design and mathematical modeling. *Advanced Drug Delivery Reviews*, Vol.58, No.12-13, (November 2006), pp. 1379-1408, ISSN 0169-409X

Lin, Y.-K; Lin, T-Y & Su, H.P. (2011). Extraction and characterisation of telopeptide-poor collagen from porcine lung, *Food Chemistry*, Vol.124, No.4, (February 2011), pp.1583-1588, ISSN 0308-8146

Liu, Y.; Bharadwaj, S.; Lee, S.J.; Atala, A. & Zhang, Y. (2009). Optimization of a natural collagen scaffold to aid cell-matrix penetration for urologic tissue engineering. *Biomaterials,*Vol.30, No.23-24, (August 2009), pp. 3865–3873, ISSN 0142-9612

Lode, A.; Bernhardt, A. & Gelinsky, M. (2008). Cultivation of human bone marrow stromal cells on three-dimensional scaffolds of mineralized collagen: influence of seeding density on colonization, proliferation and osteogenic differentiation. *Journal of Tissue Engineering and Regenenerative Medicine,* Vol.2, No.7, (October 2008), pp. 400–407, ISSN 1932-7005

Lungu, A.; Albu, M.G. & Trandafir, V. (2007). New biocomposite matrices structures based on collagen and synthetic polymers designed for medical applications. *Materiale Plastice*, Vol.44, No.4, pp. 273-277, ISSN 0025-5289

Magnan, M.; Berthod, F.; Champigny, M.F.; Soucy, F. & Bolduc, S. (2006). *In vitro* reconstruction of a tissue-engineered endothelialized bladder from a single porcine biopsy. *Journal of Pediatric Urology*, Vol.2, No.4, (August 2006), pp. 261–270, ISSN 1477-5131

McDaniel, D.P.; Gordon, A.; Shaw, J.; Elliott, T.; Bhadriraju, K.; Meuse, C.; Chung, K-H. & . Plant, A.L. (2007). The Stiffness of Collagen Fibrils Influences Vascular Smooth Muscle Cell Phenotype. *Biophysical Journal*, Vol.92, No.5, (March 2007), pp. 1759-1769, ISSN 0006-3495

Metzmacher, I.; Radu, F.; Bause, M.; Knabner, P. & Friess, W. (2007). A model describing the effect of enzymatic degradation on drug release from collagen minorods. *European Journal of Pharmaceutics and Biopharmaceutics*, Vol.67, No.2, (September 2007), pp. 349-360, ISSN 0939-6411

Minakuchi, Y.; Takeshita, F.; Kosaka, N.; Sasaki, H.; Yamamoto, Y.; Kouno, M.; Honma, K.; Nagahara, S.; Hanai, K.; Sano, A.; Kato, T.; Terada, M. & Ochiya T. (2004).

Atelocollagen-mediated synthetic small interfering RNA delivery for effective gene silencing *in vitro* and *in vivo*. *Nucleic Acids Research*, Vol.32, No.13, (July 2004), pp. e109, ISSN 0305-1048

Mooney, D.J. & Mikos, A.G. (1999). Growing new organs. *Scientific American*, Vol.280 (April 1999) pp. 60–65, ISSN 0036-8733

Nagai, T. & Suzuki, N. (2002). Preparation and partial characterization of collagen from paper nautilus (*Argonauta argo*, Linnaeus) outer skin, *Food Chemistry*, Vol.76, No.2, (February 2002), pp.149-153, ISSN 0308-8146

Natu, M.V.; Sardinha, J.P.; Correia, I.J. & Gil, M.H. (2007). Controlled release gelatine hydrogels and lyophilisates with potential applications as ocular inserts. *Biomedical Materials*, Vol.2, No.4, (December 2007), pp. 241-249, ISSN 1748-6041

Parker, D.M.; Armstrong, P.J; Frizzi, J.D & North Jr, J.H. (2006). Porcine Dermal Collagen (Permacol) for Abdominal Wall Reconstruction, *Current Surgery*, Vol.63, No.4, (July-August 2006), pp.255-258, ISSN 0149-7944

Parshotam Kumar, G.; Barker, A.; Ahmed, S.; Gerath, J. & Orford, J. (2010). Urinary bladder auto augmentation using INTEGRA((R)) and SURGISIS ((R)): An experimental model. *Pediatric Surgurgery Inernational*, Vol.26, No.3, (March 2010), pp. 275–280, ISSN 1437-9813

Peppas, N.A.; Bures, P.; Leobandung, W. & Ichikawa, H. (2000). Hydrogels in pharmaceutical formulations. *European Journal of Pharmaceutics and Biopharmaceutics*, Vol.50, No.1, (July 2000), pp. 27-46, ISSN 0939-6411

Phaechamud, T. & Charoenteeraboon, J. (2008). Antibacterial activity and drug release of chitosan sponge containing doxycycline hyclate. *AAPS PharmSciTech*, Vol.9, No.3 , (September 2008), pp. 829-835, ISSN 1530-9932

Piez, K.A. (1984). Molecular and aggregate structures of the collagens, In: *Extracellular Matrix Biochemistry*, Piez, K.A. & Reddi, A.H., pp.1-40, Elsevier, ISBN 978-0-4440-0799-5, New York

Radu, F.A.; Bause, M.; Knabner, P.; Friess, W. & Metzmacher, I.. (2009). Numerical simulation of drug release from collagen matrices by enzymatic degradation. *Computing and visualization in science*, Vol.12, No.8, pp. 409-420, ISSN 1432-9360

Rafat, M.; Matsuura, T.; Li, F. & Griffith, M. (2009). Surface modification of collagen-based artificial cornea for reduced endothelialization. *Journal of Biomedical Materials Research Part A*, Vol.*88*, (March 2009), pp. 755–768, ISSN 1549-3296

Ranucci, C.S.; Kumar, A.; Batra, S.P. & Moghe, P.V. (2000). Control of hepatocyte function on collagen foams: sizing matrix pores toward selective induction of 2-D and 3-D cellular morphogenesis. *Biomaterials*, Vol.21, (April 2000), pp.783-793, ISSN 0142-9612

Rao, P. K. (1996). Recent developments of collagen-based materials for medical applications and drug delivery systems. *Journal of Biomaterials Science, Polymer Edition*, Vol.7, No.7, (January 1996), pp. 623-645(23), ISSN 1568-5624

Raub, C.B.; Suresh, V.; Krasieva, T.; Lyubovitsky, J.; Mih, J.D., Putnam, A.J.; Tromberg, B.J. & George, S.C. (2007). Noninvasive assessment of collagen gel microstructure and mechanics using multiphoton microscopy. *Biophysical Journal*, Vol.92, No.6, (March 2007), pp. 2212-2222, ISSN 0006-3495

Renou, J.-P.; Foucat, L.; Corsaro, C.; Ollivier, J.; Zanotti, J.-M. & Middendorf H.D. (2004). Dynamics of collagen from bovine connective tissues, *Physica B: Condensed Matter*, Vol.350, No.1-3, (July 2004), pp.E631-E633, ISSN 0921-4526

Rodziewicz-Motowidło, S.; Śladewska, A.; Mulkiewicz, E.; Kołodziejczyk, A.; Aleksandrowicz, A.; Miszkiewicz, J. & Stepnowski, P. (2008), Isolation and characterization of a thermally stable collagen preparation from the outer skin of the silver carp *Hypophthalmichthys molitrix*, *Aquaculture*, Vol. 285, No.1-4, (December 2008), pp.130-134, ISSN 0044-8486

Roreger, M. (November 1995). *Collagen preparation for the controlled release of active substances*, Patent No. PCT/EP1995/001428, Germany

Rubinstein, M. P. (2003). Applications of contact lens devices in the management of corneal disease. *Eye* Vol.17, (February 2003), pp. 872–876, ISSN 0950-222X

Ruszczak, Z. & Friess, W. (2003). Collagen as a carrier for on-site delivery of antibacterial drugs. *Advanced Drug Delivery Reviews*, Vol.55, No.12, (November 2003), pp. 1679-1698, ISSN 0169-409X

Ruszczak, Z. (2003). Effect of collagen matrices on dermal wound healing. *Advanced Drug Delivery Reviews* (November 2003) Vol.55, No.12, pp. 1595-1611, ISSN 0169-409X

Sahoo, S.; Cho-Hong, J.G. & Siew-Lok, T. (2007). Development of hybrid polymer scaffolds for potential applications in ligament and tendon tissue engineering. *Biomedical Materials*, Vol.2, No.3, (September 2007), pp. 169-173, ISSN 1748-6041

Sang, L.; Luo, D.; Xu, S.; Wang, X. & Li, X. (2011). Fabrication and evaluation of biomimetic scaffolds by using collagen–alginate fibrillar gels for potential tissue engineering applications, *Materials Science and Engineering: C*, Vol.31, No.2, (March 2011), pp.262-271, ISSN 0928-4931

Satish, C.S.; Satish, K.P. & Shivakumar, H.G. (2006). Hydrogels as controlled drug delivery systems: synthesis, crosslinking, water and drug transport mechanism. *Indian Journal of Pharmaceutical Sciences*, Vol.68, No.2, (March 2006), pp. 133-140, ISSN 0250-474X

Schnell, E.; Klinkhammer, K., Balzer, S., Brook, G., Klee, D., Dalton, P. & Mey, J. (2007). Guidance of glial cell migration and axonal growth on electrospun nanofibers of poly-ε-caprolactone and a collagen/poly-ε-caprolactone blend, *Biomaterials*, Vol.28, No.19, (July 2007), pp.3012-3025, ISSN 0142-9612

Schulz, R.M.; Zscharnack, M.; Hanisch, I.; Geiling, M.; Hepp, P. & Bader, A. (2008). Cartilage tissue engineering by collagen matrix associated bone marrow derived mesenchymal stem cells. *Bio-medical materials and engineering*, Vol.18, No.1, (November 2008), pp. S55-70, ISSN: 1422-6405

Service, R.F. (2005). Tissue engineering. Technique uses body as 'bioreactor' to grow new bone. *Science*, Vol.309 No.5735, (July 2005), pp.683, ISSN 0036-8075

Singh, M.P.; Lumpkin, J.A. & Rosenblatt, J. (1995). Effect of electrostatic interactions on polylysine release rates from collagen matrices and comparison with model predictions. *Journal of Controlled Release*, Vol.35, No.2-3, (August 1995), pp. 165-179, ISSN 0168-3659

Singh, P.; Benjakul, S.; Maqsood, S. & Kishimura, H., Isolation and characterisation of collagen extracted from the skin of striped catfish (*Pangasianodon hypophthalmus*), *Food Chemistry*, Vol.124, No. 1, (January 2011), pp.97-105, ISSN 0308-8146

Sionkowska A.; Wisniewski, M; Skopinska, J., Kennedy, C.J. & Wess, T.J. (2004). The photochemical stability of collagen–chitosan blends, Journal of Photochemistry and Photobiology A: Chemistry, Vol.162, No. 2-3, (March 2004), pp.545-554, ISNN 1010-6030

Sionkowska, A.; Skopinska-Wisniewska, J., Wisniewski, M., Collagen–synthetic polymer interactions in solution and in thin films, Journal of Molecular Liquids, Vol.145, No.3, (May 2009), pp.135-138, ISSN 0167-7322

Skopinska-Wisniewska, J.; Sionkowska, A.; Kaminska, A.; Kaznica, A.; Jachimiak, R. & Drewa, T. (2009). Surface characterization of collagen/elastin based biomaterials for tissue regeneration, Applied Surface Science, Vol.255, No. 19, (July 2009), pp.8286-8292, ISSN 0169-4332

Smith, M.; McFetridge, P.; Bodamyali, T.; Chaudhuri, J.B.; Howell, J.A.; Stevens, C.R. & Horrocks, M. (2000). Porcine-Derived Collagen as a Scaffold for Tissue Engineering, Food and Bioproducts Processing, Vol.78, No.1, (March 2000), pp.19-24, ISSN 0960-3085

Stadlinger, B.; Pilling, E.; Huhle, M.; Mai, R.; Bierbaum, S.; Scharnweber, D.; Kuhlisch, E.; Loukota, R. & Eckelt, U. (2008) Evaluation of osseointegration of dental implants coated with collagen, chondroitin sulphate and BMP-4: an animal study, International Journal of Oral and Maxillofacial Surgery, Vol.37, No.1, (January 2008), pp. 54-59, ISSN 0882-2786

Stamov, D.; Grimmer, M.; Salchert, K.; Pompe, T. & Werner, C. (2008). Heparin intercalation into reconstituted collagen I fibrils: Impact on growth kinetics and morphology. Biomaterials, Vol.29, No.1, (January 2008), pp.1-14, ISSN 0142-9612

Stark, Y.; Suck, K.; Kasper, C.; Wieland, M.; van Griensven, M. & Scheper, T. (2006). Application of collagen matrices for cartilage tissue engineering. Experimental and Toxicologic Pathology, Vol.57, No.4, (March 2006), pp. 305-311, ISSN: 0940-2993

Stojadinovic, A.; Carlson, J.W.; Schultz, G.S.; Davis, T.A. & Elster, E.A. (2008). Topical advances in wound care. Gynecologic Oncology, Vol.111, No.3, Suppl.1, (November 2008), pp. S70-S80, ISSN 0090-8258

Su, S.-C.; Mendoza, E.A.; Kwak, H. & Bayless, K.J. (2008). Molecular profile of endothelial invasion of three-dimensional collagen matrices: insights into angiogenic sprout induction in wound healing. American Journal of Physiology - Cell Physiology, Vol. 29, (September 2008), pp.C1215–C1229, ISSN 1522-1563

Sun, W.; Lin, H.; Chen, B.; Zhao, W.; Zhao, Y. & Dai, J. (2007). Promotion of peripheral nerve growth by collagen scaffolds loaded with collagen-targeting human nerve growth factor-beta. Journal of Biomedical Materials Research Part A, Vol.83, No.4, (December 2007), pp. 1054–1061, ISSN 1549-3296

Sun, W.; Lin, H.; Chen, B.; Zhao, W.; Zhao, Y.; Xiao, Z. & Dai, J. (2009). NGF-beta accelerate ulcer healing. Journal of Biomedical Materials Research Part A, Vol.92A, No.3, (March 2009), pp.887–895, ISSN 1549-3296

Swatschek, D.; Schatton, W.; Müller, W.E.G. & Kreuter, J. (2002), Microparticles derived from marine sponge collagen (SCMPs): preparation, characterization and suitability for dermal delivery of all-trans retinol, European Journal of Pharmaceutics and Biopharmaceutics, Vol.54, No.2, (September 2002), pp.125-133, ISSN 0939-6411

Tebb, T.A.; Tsai, S.W.; Glattauer, V.; White, J.F.; Ramshaw, J.A. & Werkmeister, J.A. (2006). Development of porous collagen beads for chondrocyte culture. *Cytotechnology*, Vol.52, No.2, (October 2006), pp. 99–106, ISSN 0920-9069

Teles, H.; Vermonden, T.; Eggink, G.; Hennink, W.E. & de Wolf, F.A. (2010). Hydrogels of collagen-inspired telechelic triblock copolymers for the sustained release of proteins. *Journal of Controlled Release*, Vol.147, No.2, (October 2010), pp. 298-303, ISSN 0168-3659

Thacharodi D. & Rao, K.P. (1996). Rate-controlling biopolymer membranes as transdermal delivery systems for nifedipine: Development and *in vitro* evaluations. *Biomaterials*, Vol.17, No.13, (July 1996), pp. 1307-1311, ISSN 0142-9612

Tillman, B.W.; Yazdani, S.K.; Lee, S.J; Geary, R.L; Atala, A. & Yoo, J.J. (2009). The *in vivo* stability of electrospun polycaprolactone-collagen scaffolds in vascular reconstruction. *Biomaterials*, Vol.30, No.4, (February 2009), pp.583-588, ISSN 0142-9612

Titorencu, I.; Albu, M.G.; Giurginca, M.; Jinga, V.; Antoniac, I.; Trandafir, V.; Cotrut, C.; Miculescu, F. & Simionescu, M. (2010). In *Vitro* Biocompatibility of Human Endothelial Cells with Collagen-Doxycycline Matrices. *Molecular Crystals and Liquid Crystals*, Vol.523, (May 2010), pp. 82/[654] - 96/[668], ISSN 1542-1406

Trandafir, V.; Popescu, G.; Albu, M.G.; Iovu, H. & Georgescu, M. (2007). *Collagen based products*, Ars Docendi, ISBN 978-973-558-291-3, Bucharest, Romania

Trelstad, R.L. (1982). Immunology of collagens, In: *Immunochemistry of the extracellular matrix*, Furthmayer, H., pp.32-39, CRC, ISBN 978-0-8493-6196-8, Boca Raton, Florida

Ueno, Y.; Nagai, H.; Watanabe, G.; Ishikawa, K.; Yoshikawa, K.; Koizumi, Y.; Kameda, T. & Sugiyama, T. (2005). Transplantation of rat hepatic stem-like (HSL) cells with collagen matrices. *Hepatology Research*, Vol.33, No.4, (December 2005), pp. 277-84, ISSN 1386-6346

Uneo, Y.; Terada, K.; Nagai, H.; Watanabe, G.; Ishikawa, H.; Yoshikawa, K.; Hirayama, Y.; Endo, E.; Kameda, T.; Liu, Y. & Sugyiama, T. (2004). Application of collagen scaffolds for hepatic stem-like cells transplantation. *Akita Journal Medicine*, Vol.31, No.2, (January 2004) pp. 113-119, ISSN 0386-6106

Uriarte-Montoya, M.H; Arias-Moscoso, J.L.; Plascencia-Jatomea, M.; Santacruz-Ortega, H.; Rouzaud-Sández, O.; Cardenas-Lopez, J.L.; Marquez-Rios, E., Ezquerra-Brauer, J.M., Jumbo squid (*Dosidicus gigas*) mantle collagen: Extraction, characterization, and potential application in the preparation of chitosan–collagen biofilms, *Bioresource Technology*, Vol.101, No.11, (June 2010), pp.4212-4219, ISSN 0960-8524

Wallace, D. G. & Rosenblatt, J. (2003). Collagen in drug delivery and tissue engineering. *Advanced Drug Delivery Review*, Vol.55, No.12, (November 2003), pp. 1631-1649, ISSN 0169-409X

Wang, D-A.; Williams, C.G.; Yang, F. & Elisseeff, H. (2004). Enhancing the tissue-biomaterial interface: tissue-initiated integration of biomaterials, *Advanced Functional Materials*, Vol.14, No.12, (December 2004), pp.1152-1159, ISSN 1616-3028

Wang, L.; An, X.; Yang, F.; Xin, Z.; Zhao, L. & Hu, Q. (2008). Isolation and characterisation of collagens from the skin, scale and bone of deep-sea redfish (Sebastes mentella), *Food Chemistry*, Vol.108, No.2, (May 2008), pp.616-623, ISSN 0308-8146

Wang, Y.; Zhang, L.; Hu, M.; Wen, W.; Xiao, H. & Niu, Y. (2010). Effect of chondroitin sulfate modification on rhBMP-2 release kinetics from collagen delivery system. *Journal of*

Biomedical Materials Research Part A, Vol.92A, No. 2, (February 2010), pp. 693–701, ISSN 1549-3296

Weiner, A.L.; Carpenter-Green, S.; Soehngen, E.C.; Lenk, R.P. & Popescu, M.C. (1985). Liposome–collagen gel matrix: A novel sustained drug delivery system. *Journal of Pharmaceutical Sciences,* Vol.74, No.9, (September 1985), pages 922–925, ISSN 0022-3549

Wen, F.; Chang, S.; Toh, Y.C.; Teoh, S.H. & Yu, H. (2007). Development of poly (lactic-co-glycolic acid)-collagen scaffolds for tissue engineering, *Materials Science and Engineering: C,* Vol.27, No.2, (March 2007), pp. 285-292, ISSN 0928-4931

Williams, D.F. (May 1999). *The Williams Dictionary of Biomaterials,* Liverpool University Press, ISBN 978-0-8532-3921-5, Liverpool, United Kindom

Wollina, U.; Meseg, A. & Weber, A. (2011), Use of a collagen–elastin matrix for hard to treat soft tissue defects. *International Wound Journal,* Vol.8, (March 20110), pp. 1742-4801, ISSN 1742-4801

Woodard, J.R.; Hilldore, A.J.; Lan, S.K.; Park, C.J.; Morgan, A.W.; Eurell, J.A.; Clark, S.G.; Wheeler, M.B.; Jamison, R.D. & Wagoner, J.A.J. (2007). The mechanical properties and osteoconductivity of hydroxyapatite bone scaffolds with multi-scale porosity. *Biomaterials,* Vol.28, No.1,(August 2007), pp. 45–54, ISSN 0142-9612

Yarboro, S.R.; Baum, E.J. & Dahners, L.D. (2007). Locally administered antibiotics for prophylaxis against surgical wound infection. *The Journal of Bone and Joint Surgery,* Vol.89, No.5, (May 2007), pp. 929-933, ISSN 0021-9355

Yoel, G. & Guy, K. (2008). Use of collagen shields for ocular-surface drug delivery. *Expert Review of Ophthalmology,* Vol.3, No.6, (December 2008), pp. 627-633(7), ISSN 1746-9899

Yu, X.; Bichtelen, A.; Wang, X.; Yan, Y.; Lin, F.; Xiong, Z; Wu, R.; Zhang, R. & Lu, Q. (2005). Collagen/Chitosan/Heparin Complex with Improved Biocompatibility for Hepatic Tissue Engineering. *Follow Journal of Bioactive and Compatible Polymers,* Vol.20, No.1, (January 2005), pp. 15-28, ISSN 0883-9115

Zhang, L.; Tang, P.; Xu, M.; Zhang, W.; Chai, W. & Wang, W. (2010). Effects of crystalline phase on the biological properties of collagen–hydroxyapatite composites, *Acta Biomaterialia,* Vol.6, No.6, (June 2010), pp. 2189-2199, ISSN 1742-7061

Zhou, J.; Cao, C.; Ma, X. & Lin, J. (2010). Electrospinning of silk fibroin and collagen for vascular tissue engineering, *International Journal of Biological Macromolecules,* Vol.47, No. 4, (November 2010), pp. 514-519, ISSN 0141-8130

Zilberman, M.& Elsner, J.J. (2008). Antibiotic-eluting medical devices for various applications. *Journal of Controlled Release,* Vol.130, No.3, (September 2008), pp. 202-215, ISSN 0168-3659

6

Biopolymers as Wound Healing Materials: Challenges and New Strategies

Ali Demir Sezer[1] and Erdal Cevher[2]
[1]Faculty of Pharmacy, Marmara University,
[2]Faculty of Pharmacy, Istanbul University,
Turkey

1. Introduction

Wound healing is a multi-factorial physiological process. The complexity of this phenomenon makes it prone to several abnormalities. Apart from cellular and biochemical components, several enzymatic pathways also become active during repair and help the tissue to heal. The goal of this chapter is to introduce the biomaterials community to the emerging field of self-healing materials, and also to suggest how one could utilize and modify self-healing approaches to develop new classes of biomaterials. On the other hand, natural and synthetic gel-like materials, films/membranes, composites, micro-/nanoparticulate systems have featured heavily in the development of biomaterials for wound healing and other tissue-engineering purposes. Nanofibrous membranes are highly soft materials with high surface-to-volume ratios, and therefore can serve as excellent carriers for therapeutic agents or accelerate wound healing. Biocompatible and biodegradable polymer scaffolds combined with cells or biological signals are being investigated as alternatives to traditional options for tissue reconstruction and transplantation. These approaches are already in clinical use as engineered tissues that enhance wound healing and skin regeneration. This chapter covers the recent reports on the preparation and biomedical applications of biopolymers and biomaterials based on pharmaceutical dosage forms and wound dressing.

2. Skin structure

Skin is the largest organ of the integumentary system, comprising 15 % of the body weight. It protects the organism against injury and damage, and prevents passing of microorganisms and lets water vapour permeation. Substances secreted by glands in the epidermis contribute to the preservation of water - electrolyte balance of the body. The skin regulates body temperature by vasodilation and vasoconstriction of cutaneous blood vessels, and is formed by two main layers, including epidermis covering the body surface, and the dermis involving the connective tissue (Junqueria et al, 1992).

2.1 Epidermis

The epidermis consists of two different cell layers. It comes into existence whereas ectoderm is taking its form. Cells forming the top layer of epithelium undergo keratinisation and form the dead layer of skin. Whereas dead cells are being pushed from the deeper portion of the epidermis toward the surface, they are replaced by proliferating cells by mitosis in the basal layer. This change is called cytomorphosis. Cytomorphosis takes an average of 15 to 30 days for a healthy individual. The basic cell type of epidermis is keratinocytes. The content of the keratin constantly changes in the epidermis because keratinocytes undergo changes in epidermis and consists of 85 % of total protein of the *Stratum corneum* (Gartner et al., 2001). Keratinocytes synthesize different keratins in different stages of cytomorphosis. Whereas basal layer cells include only low molecular weight keratins, cells with high molecular weight keratin begin to build the heap structure of the *Stratum corneum* as the cells move upward. Keratins in this layer are cross-connected by disulphide bonds at the same time.

The epidermis consists of five layers. These are, respectively, from the inside to out (Junqueria et al., 1992).

Stratum basale (germinativum): This layer consists of a single row of cylindrical cells sitting on the basal membrane, located on the dermis and basal lamina. This layer is responsible for continuous renewal of the epidermis due to the proliferation of cells. Cytoplasmic fibrillar protein is found in all cells that form the structure. As proliferating cells migrate towards the upper layer, the number of protein filaments increases.

Stratum spinosum: This is a multi-storey layer composed of polygonal or squamous epithelial cells. The cytoplasm and cytoplasmic extensions are filled with bundles of fibrils and connected to each other with bridges on the cell surface. These cells are called spino-cellular cells, due to the spine-like structures around this layer of cells. All the layers of both *Stratum basale* and *Stratum spinosum* are called the Malpighian layer.

Stratum granulosum: This layer consists of small, flat, polygonal-shaped cells. These cells contain keratohyalin granules. The number of lamellar structures emerging in the *Stratum spinosum* increases in the *Stratum granulosum*, their contents are secreted to intercellular spaces as accumulating in the periphery of the cell, providing intercellular lipids, which are essential for the barrier function of the epidermis.

Stratum lucidum (Transparent layer): This layer is composed of flat, transparent, and tightly gathered cells ranging from 3 to 5 rows. It is a thin layer that is only found in regions where the epidermis is thick. The cytoplasm of the *transparent layer* is filled with a substance refracting light called eleidin and cell organelles decrease in this layer. As tonofibrils become more numerous and regular, they make the cell membrane thicker. Desmosomes are located between cells and the amount of intercellular materials increases.

Stratum corneum: It is a dead cell layer having no nucleus. Keratin is found in the cytoplasm of these cells. Intercellular gaps are full of lipids which are secreted from the lamellar structures in the *Stratum spinosum* and *Stratum granulosum*. Superficial cells of the *Stratum corneum* shed continuously (desquamation) and new tissue is produced by mitosis of cells in the germinal layer. Keratinization occurs with the formation of disulfide groups from sulphydryl groups of protein fibrils during migration to the upper layer. These protein fibrils make reticulated bundles, forming a substantial long chain by disulfide bonds and also including a dense, amorphous matrix between them, composed of keratohyalin granules. At this stage, the cell membrane thickens. This layer, formed by the thickening of the cell membrane, takes the shape of cornified cells losing its core and other organelles after reaching of keratohyalin granules to maximum point and forming of keratin lipids (Gartner et al., 2001; Junqueria et al., 1992).

2.2 Dermis

The dermis is a type of elastic, flexible bond tissue locating under epidermis and is vascularised, enabling it to provide energy and nutrition to the epidermis. The dermis extends limitlessly by integrating to the subcutaneous layer and its thickness varies according to the region (Leeson & Leeson, 1981). Its surface contacting the epidermis is rough and composed of papillae. Papillae are named *Stratum papillare* by integrating deep surface of the epidermis. Some papillae also contain special nerve endings, called vascular papillae (Young & Heath, 2000). The deepest part of dermis is called as *Stratum reticulare*. This layer consists of irregular, dense connective tissue and has rather weak cells, primarily fibroblasts and macrophages. Hair follicles are concentrated around sebaceous and sweat glands in the dermis (Leeson & Leeson, 1981).

2.3 Hypodermis

Hypodermis, the subcutaneous layer, is a loose connective tissue located under the dermis and containing varying amounts of flat cells. Collagen and elastic fibrils within its structure continue into the dermis. Hair roots also are found in this layer (Junqueria et al., 1992; Leeson et al., 1988). The increase in the number of fibrils results in a rigid binding of dermis to hypodermis and thus affects the mobility of the skin.

3. Wounds and burns

3.1 Types of wounds

A wound is the disruption of the integrity of anatomical tissues caused by exposure to any factor. Wounds are examined under two groups:

Closed Wounds: This group includes contusion, hematoma and abrasion. Contusion-type injuries involve damage to soft tissues, small blood vessels and deep tissue layers, resulting in their separation, but the anatomy of the skin remains intact. Oedema, and in later periods, atrophy and defective pigmentation are observed in wound and the healing is delayed. Vessel rupture or hyperaemia due to vessel damage is called hematoma and wounds such as scrapes are termed abrasions. The healing process is very painful because this type of wound involves damage to sensory nerves and the wound can easily become infected (Mutsaers et al., 1997).

Open Wounds: This group includes lacerations, cutting-pricking tool wounds, gunshot wounds, surgical wounds, insect bites and stings, radionecrosis, vascular neurological and metabolic wounds. Wounds except for lacerations cause serious damage to tissues beneath the skin. In laceration type wounds, skin and subcutaneous tissue have been destroyed, but deep tissues remain healthy. The anatomical integrity of tissues is damaged in cutting-pricking tool wounds without any tissue damage at the edges of the wound (Aydın, 2000; Kapoor & Appleton, 2005).

Wounds are also classified according to tissue loss.

Wounds with Tissue Loss: These types of wounds involve damage or loss in some or all of the skin layers. Healing occurs *via* filling of the wound area by granulation tissue typically growing from the base of a wound. Wounds that involve tissue loss are collected in two groups in proportion to the loss. In superficial wounds, the entire epidermis and the papillar layer of the dermis are damaged. The epidermis, all the layers of the dermis and even subcutaneous tissue are damaged in full-thickness wounds covering second group (Mutsaers et al., 1997; Porth, 1998; Chanson et al., 2005).

Wounds without Tissue Loss: These kinds of wounds occur as a result of tissue crushing. The severity of bleeding occurring in tissue varies according to the condition of the wound. Tissues exposed to this kind of wound heal after granulation tissue formation in minimal level in first phase of healing process (Ruszczak, 2003).

3.2 Types of burns

Burn is a kind of wound that occurs when skin or organs are damaged by an electrical current, heat, chemical or flammable agent effect. It is known that burn-initiated pathophysiological events differ from other traumas, may cause death risks and lead to increasing capillary permeability, resulting in hypovolemia. Burn causes changes of vascular permeability, extravasation of plasma proteins, aggregation of platelets and increased fibrinolysis (Yenerman, 1986; Madri, 1990; Atiyeh et al., 2005).

Burns are divided into 4 groups according to the depth and the affected skin layers:

3.2.1 First-degree burns

Only the outer layer of the epidermis and *Stratum corneum* are damaged in this type of burn, and there is no damage in the dermis. First-degree burns generally occur as a result of short-term heat or flame contact or long-term exposing to intense sunlight. First-degree burn areas are characterised by slight oedema, which diminishes after 24 hours. At this stage the skin begins to dry, there is no vesicle and infection is not seen. The wound heals within a week (Whitney & Wickline, 2003).

3.2.2 Second-degree burns

These types of burns are deeper than first-degree burns and necrosis spread into the dermis. Damage covers the entire epidermis and some part of dermis. The wound is clinically characterised by pain, erythema and bullae. The recovery rate depends on the depth of skin injury and formation of infection. Generally, second-degree burns heal spontaneously in a short period if infection does not occur. If infection occurs in the wound, it can easily convert to third degree burn.

The burns in this group may be divided into two categories, termed superficial and deep dermal second-degree burns (Sparkes, 1997; Whitney & Wickline, 2003):

Superficial Second Degree Burns: These occur due to short period contact with flame or hot liquids. Generally, the upper portion of the *Stratum germinativum* is damaged in superficial second-degree burns. The surface is generally humid because of the leakage of liquid plasma from the burned area. Generally, less scarring than the deep dermal burns occurs. Recovery usually occurs within 3-4 weeks with zero or very mild scarring (Madri, 1990).

Deep Dermal Burns: These occur due to contact with chemicals such as flame, hot liquids or acids, or exposure to high electrical current. In these types of burns, whereas the epidermis is completely burned, damage extends to the *Stratum germinativum* and the bottom section of the dermis. In deep dermal burns, fluid loss and metabolic effects are the same as in third-degree burns. Wound pain is very severe during burning and hyperanesthesia may occur in some areas. The wound may develop into a third-degree burn if infection occurs in the burn area. The time required for re-epithelialisation depends on degradation in the dermis, the amount of burnt hair follicles and sweat glands, and the width of infected areas. If the wound is properly preserved, it usually closes within 2 months, leaving some scarring on the skin surface. Scarring and contracture occur in healing areas where closing lasts longer than 2 months. In this case, it is very difficult to distinguish the wound from third-degree burns and treatment of this kind of wound will last longer (Porth, 1998; Shakespeare, 2001).

3.2.3 Third-degree burn

These kinds of burns result from hot water, fire and prolonged contact with electrical current. The wound area exhibits tautness and brightness as the elasticity of the skin is lost, causing abnormal shrinkage. In such cases, all structures within the skin sustain damage. The dermis and subcutaneous fat are destroyed as a result of coagulation necrosis. Thrombosis occurs in vessels under the skin. Increased capillary permeability and oedema is much higher in third-degree burns than second-degree burns. Skin is damaged in all layers and is characterised by autolysis and leukocytoclastic infiltration for 2 or 3 weeks. This event is usually associated with suppuration. Capillary bundles and fibroblasts are organised in granulation tissue under scar. If the burn affects subcutaneous fat, healing can take much longer. Burn affecting the muscle causes increasing in degradation of red blood cells. The care of third-degree burns requires removing scar tissue and covering the wound with a graft. If grafting is not carried out, a thick layer of granulation is shaped and the contraction of the area follows it. At this stage, re-epithelialisation, slightly, occurs on the edge of wound granulation is soft, can be infected and healing continues over several months. Permanent deep scars in the skin occur following healing in these kinds of wounds and surgical intervention is usually required to restore normal appearance (Moulin et al., 2000; Shakespeare, 2001).

3.2.4 Fourth-degree burns

This refers to the carbonization of burned tissues.

4. Wound and burn treatment

4.1 Developments in the treatment of wounds and burns

A range of methods has been used to treat wounds, dating back to ancient times. The earliest information on wound treatment is found in Egyptian medical documents, called the Ebers Papyrus. It is known that ancient Egyptians treated wounds by covering them frog skin and castor oil. The results have been limited and partially misleading, although humans have used many materials of biological origin in wound and burn treatment throughout history and have conducted various experiments on animals. The first wide-ranging microscopic study was conducted by Hartwell in the 1930s. Hartwell compared human wounds to those in pigs, rabbits, dogs and guinea pigs and found that the wound healing progress is different in human epithelial and subepithelial surfaces, compared to those of animals. If the pathological table of wound healing is also taken into account, it was confirmed that pig physiopathology was most similar to humans, followed by that of rabbits. Later, Gangjee et al. (1985) conducted studies of percutaneous wound healing. Winter et al. (1965) developed animal models of wounds and burns using pigs as a subject for wound treatment and concluded that, as the histology of skin and the wound healing mechanism differs between animals and humans, animal models or experiments can only provide a general indication of wound healing phases.

Researchers subsequently sought new materials for use in wound healing, due to the disadvantages of traditional dressings such as gauze, paraffin gauze, biological dressings, etc. The first synthetic material used for wound coverage was methyl cellulose. The common feature of all these materials is the necessity of physical protection from external factors and conditions. Many new materials have been used as the wound healing mechanism has become better understood. The usage of artificial dressing materials in forms such as film, spray, foam and gel have increased significantly in recent years (Shakespeare & Shakespeare, 2002; Stashak et al., 2004; Merei, 2004)

The following characteristics are required for ideal wound and burn dressing (Sheridan & Tompkins, 1999; Balasubramani et al., 2001; Jones et al., 2002);

- ease of application
- bioadhesiveness to the wound surface
- sufficient water vapour permeability
- easily sterilised
- inhibition of bacterial invasion
- elasticity and high mechanical strength
- compatibility with topical therapeutic agents
- optimum oxygen permeability
- biodegradability
- non-toxic and non-antigenic properties

Average water loss from normal human skin is 250 $g/m^2/day$. In wounded skin, this figure can reach up 5000 $g/m^2/day$ according to the type of wound. If a wound dressing material is thin and has extreme water vapour permeability, it causes accumulation of liquid, bacterial growing and delay in recovery (Alper et al., 1982; Fansler et al., 1995). Wound coverings should be adequately adhesive to the wound surface and edges. At the end of treatment, cover material should also permit removal in such a manner that it does not harm any tissues, must not cause toxic or antigenic reactions, and must be biocompatible. The materials used should not cause contamination of microorganisms on the wound surface and, if possible, it should also prevent the proliferation of bacteria in normal skin flora. Wound dressing materials should be easy to apply and have sufficient elasticity and mechanical strength to be used in areas especially close to the joints. The surface of the dressing material that is in contact with the wound should support the development of fibrovascular tissue as creating a convenient platform for wound healing. Adequate oxygenation is important in wound healing, so dressing material should provide contact the wound with oxygen. In addition, major desirable characteristics are that wound dressing materials should be easily sterilised, have a long shelf-life, and be cost-effective (Quinn et al., 1985; Lloyd et al., 1998; Stashak et al., 2004).

4.2 Classification of dressings used in wound and burn treatment

Materials used to cover wounds and burns are also called artificial skin, as they fulfil the functions of normal skin within areas with wounds and partly destroyed skin.

Wound and burn covering materials are classified as follows (Freyman et al., 2001; Stashak et al., 2004);

1. Traditional dressing
2. Biomaterial-based dressings
3. Artificial dressings

4.2.1 Traditional dressing

These are still the most commonly used materials for wound and burn dressings (Balasubramani et al, 2001). The traditional dressings, which are generally used during first intervention in wound treatment, prevent wound's contact with outer environment and bleeding (Sheridan & Tompkins, 1999; Stashak et al., 2004). The best sample of this group is gauze and gauze-cotton composites which have very high absorption capacity. As they cause rapid dehydration whereas they are being removed from the wound surface, they can cause bleeding and damage of newly formed epithelium (Naimer & Chemla, 2000; Stashak et al, 2004). Therefore, gauze composites with a non-adhesive inner surface are prepared to reduce the pain and trauma which can occur when removing traditional wound dressings from the wound surface.

Traditional wound dressings in the world market are shown in Table 1.

Dressing material	Brand name	Manufacturer
Paraffin gauze dressing containing 0.5% chlorhexidine acetate	Bactigras	Smith & Nephew
Paraffin gauze dressing	Jelonet	Smith & Nephew
Petrolatum gauze	Xeroform	Chesebrough-Pond's Inc.
Petrolatum gauze containing 3% bismuth tribromophenate	Xeroform	Chesebrough-Pond's Inc.
Scarlet Red dressing	Scarlet Red	Chesebrough-Pond's Inc.
Sterile hydrogel dressing	2nd skin®	Spenco
Highly absorbent cotton wool pad	Gamgee® pad	3M
Highly absorbent rayon/cellulose blend sandwiched with a layer of anti-shear high density polyethylene	Exu Dry Dressing	Smith & Nephew
Absorbent cotton pad	Telfa "Ouchless" Nonadherent Dressings	Kendall (Covidien)

Table 1. Traditional wound dressings in the world market.

Exudate leaking from traditional dressing materials usually increases the risk of infection and is one of the most significant problems of these type dressings. Antibacterial agents are added into the dressings to eliminate the infection. In addition, one of the most significant problems encountered in this material is foreign body reaction in the wound caused by cotton fibres. The biggest advantage of these materials is their low cost (Lim et al., 2000; Price et al., 2001; Stashak et al., 2004).

4.2.2 Biomaterial-based dressings

The most convenient method used in complete closure of wounds and burns is autografting. However, inadequate donor areas for large wounds led to the search for a new tissue source (Sheridon et al., 2001). Biological dressings are natural dressings with collagen-type structures, generally including elastin and lipid.

Such dressings can mainly be categorised under the following groups (Sheridon et al., 2001; Kearney, 2001).

1.   Allografts
2.   Tissue derivatives
3.   Xenografts

Some biomaterial-based dressings, which were used for the treatment of wounds and burns, is shown in Table 2.

Type of dressing	Dressing material	Brand name	References/ Manufacturer
Allograft	Scalp tissue	-------	Barnett et al. (1983)
	Amniotic membrane	-------	Peters and Wirth (2003)
Xsenograft	Porcine tissue	Mediskin	Genetic Lab.
	Silver impregnated porcine tissue	E-Z derm	Genetic Lab.
Skin derivatives	Highly purified bovine collagen	-------	Chvapil et al. (1973)
	Formaline fixed skin	-------	Chvapil et al. (1973)

Table 2. Biomaterial-based dressings.

Allografts

The most common source for this type of dressing is fresh or freeze-dried skin fragments taken from the patient's relatives or cadavers. Immune reaction as a result of the use of allograft can be seen and the body may reject the tissue. Infection risk also increases with suppression of the immune system to prevent the body's rejection of transplanted tissue.

The other disadvantages of these dressings include the difficulty of preparation, lack of donors, high cost and limited shelf life (Nanchahal et al., 2002; Ruszczak, 2003).

Amniotic membrane, which is separated from chorion, generally uses in superficial partial thickness burns as a dressing material for many years (Ravishanker et al., 2003). Though it has advantages such as ease of preparation and use, it has disadvantages like causing cross-infection and dehydration of the wound (Freyman et al., 2001; Jones et al., 2002). Amniotic membrane derived from a healthy donor is shown in Figure 1.

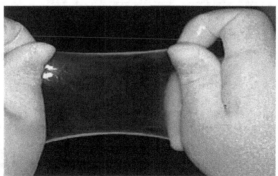

Fig. 1. The cryo-preserved amniotic membrane was thawed prior to its application (Hasegawa et al., 2007).

The effectiveness of amniotic membranes as dressing materials in burn treatment is summarized in Table 3.

Dressing quality	Amniotic membrane
Pain relief	+++
Infection prevention	+++
Good adherence to wound bed	+++
Promotion of re-epithelialisation	+++
Cost factor	++
Moist wound healing environment	++
Elasticity/conformability	+++
Easy application	++
Prevention of heat loss	+
Availability	++

+++, very good; ++, average; +, below average
Table 3. Qualities of an ideal dressing for partial-thickness burns (Branski et al., 2008).

In vitro epidermal cell cultures are new and currently expensive systems used to prepare dressing materials. These can be applied in the form of autologous or homologous epithelial cell cultures and are still at the development stage (Beckenstein et al., 2004; Manwaring et al., 2004).

Xenografts

Xenografts are commercially available materials contrary to autografts and allografts. The most common of xenografts is the ones derived from pig skin (Sheridan et al., 2001) which have a long shelf-life and can be sterilized easily (Sheridan & Tompkins, 1999). Although pig skin is not microscopically similar to human skin, it shows close similarity in terms of adhesion and collagen content. Its disadvantage is the risk of triggering an immune response due to the foreign tissue.

Tissue derivatives

These materials, derived from different forms of collagen, have the advantages like ease of preparation, low contamination risk and weak antigenic features. The greatest disadvantage of these materials is the risk of infection, particularly in long term usage (Jones et al., 2002; Stashak et al., 2004).

4.2.3 Artificial dressings

The usage of traditional dressing materials and biomaterial-based dressings is restricted due to factors such as their stability problems and risk of infection. These conditions brought up the use of wound and burn dressing materials being cheaper and more effective, and having long shelf-life. Many dressing materials have ideal features for the treatment of wounds and burns; however, due to the variations between pathophysiology of the wound and burn, it is difficult to develop an artificial dressing material that meets all the criteria for optimum healing. Much research is currently being undertaken studies to develop wound dressing materials that can provide optimum healing conditions, taking into account all of these factors and healing mechanisms (inflammation, tissue replacement, fibrosis, coagulation, etc.) (Still et al., 2003; Stashak et al., 2004).

Some of artificial wound and burn dressing materials in world market are shown in Table 4.

Type	Dressing material	Brand name	Company
Film/membrane	Polyurethane	Omiderm	Omicron Scientific
		Opsite	Smith & Nephew
		Bioclusive	Johnson & Johnson
		Tegaderm	3M
	Polyvinyl chloride	Strech Neal	Colgate
	Nylon velour	Capran77C	Allied Chemical Corp.
	Polyvinylidene chloride	Saran Wrap	Asahi Kasei
	Polyurethane hydrocolloid	Granuflex	ConvaTec
	Synthetic fibre + aluminium	Aluderm	Söhngen
	Synthetic fibre + metal	Scanpore tape	Norgeplaster
	Activated charcoal	Carbopad	Charcoal
Foam	Formalinized polyvinyl alcohol	Ivalon	Chardack
	Polyurethane	Lyofoam	Ultra Labs
	Poly(dimethylsiloxane)	Silastic	Dow Corning
Gel	Calcium alginate	Kaltostat	ConvaTec
	Polyurethane with grafted acrylamide and hydroxyethylmethacrylate	Omiderm	Omicron Scientific
Composite	Polyprophylene film and polyurethane foam	Epigard	Parke-Davis
	Silicone film with a nylon fabric	Biobrane	Smith & Nephew
Spray	Methacrylic acid ethoxyethyl ester	Nobecutane	Astra Zeneca
	Polyhydroxyethylmethacrylate and polyethylene glycol 400	Hydron	Hydron Lab.

Table 4. Artificial wound and burn dressings in world market.

4.2.3.1 Polymers used for artificial dressings

Many natural and synthetic polymers are being used in the preparation of artificial dressing materials.

The most widely used of these includes:

a. Natural polymers

Collagen

Collagen is a biodegradable and biocompatible protein mostly found in connective tissue. The first medical usage of collagen in humans was reported by Knapp et al. (1977) and was used to provide co-reaction of contour deformities. Bovine collagen was used as suture and hemostatic agents after years. In 1980, Zyderm 1 was released, a suspension form containing sterilised fibrillar bovine collagen that was used for injecting under the dermis in wounds. Today, collagen is used in numerous biomedical applications (Hafemann et al., 1999; Ortega & Milner, 2000). These include collagen suspensions for dermal injection, topical haemostatic agents, wound dressing materials, collagen suture and catguts, collagen gels for periodontal reconstruction, collagen sponges for the hemostasis and coating of joint, and collagen rich pig skin wound dressing materials (Gingras et al., 2003; Park et al., 2004).

Alginic acid and its salts

Alginic acid, is a natural polysaccharide derived from brown algae such as *Laminaria* and *Ascophyllum* species. Alginic acid are formed by linear block copolymerization of D-mannuronic acid and L-guluronic acid. Alginic acid and its salts are used for the treatment of wound and burn due to their haemostatic properties. Their first applications were in the form of a gel, but sponges produced from calcium alginate are also used effectively in the treatment of wounds. It is also indicated that calcium alginate increases cellular activity properties such as adhesion and proliferation (Thomas, 2000a, 2000b and 2000c).

Hyaluronic acid and its derivatives

Hyaluronic acid is a natural biopolymer that alternately consists of D-glucuronic acid and 2-acetamido-2-deoxy-D-glucose and is generally found in mammal's bond tissues and synovial fluids (Saliba, 2001; Kirker et al., 2002). It has been reported that hyaluronic acid interacts with proteins, proteoglycans, growth factors and tissue components called biomolecules which has vital importance in healing of various types of wounds (Park et al., 2003). This interaction plays an important role in acceleration of tissue repair and wound healing. Hyaluronic acid and its derivatives also play a role in the protection of the injured area against microorganisms due to their bacteriostatic activity (Miller et al., 2003; Lobmann et al., 2003).

Chitosan

Chitosan, which is produced by deacetylation of kitin, is a linear polysaccharide composed of randomly distributed β-(1-4)-linked D-glucosamine (deacetylated unit) and N-acetyl-D-glucosamine (acetylated unit) (Muzzarelli & Muzzarelli, 2002; Krajewska, 2004). Chitosan is used in the treatment of wounds and burns due to its haemostatic effect (Ueno et al., 1999; Khor & Lim, 2003; Şenel & McClure, 2004). It is thought that chitosan accelerates the formation of fibroblasts and increases early phase reactions related to healing (Paul & Sharma, 2004). Chitosan can be prepared in a variety of forms, namely films, hydrogels, fibres, powders and micro-/nanoparticles. Applications of this biopolymer in commercial and biomedical fields have increased due to the low toxicity of chitosan and its biodegradation products, and its biocompatibility with blood and tissues (Berthold et al., 1994; Cho et al., 1999; Tan et al., 2001; Ishihara et al., 2002; Kim et al., 2002; Mi et al., 2002).

Fucoidan

Fucoidan is a sulphated polyfucose polysaccharide and has attracted considerable biotechnological research interest since the discovery that it possessed anti-coagulant activity similar to that of heparin and also reported to possess other properties including anti-thrombotic, anti-inflammatory, anti-tumoral and anti-viral effects (Patankar et al., 1993). Many of these effects are thought to be due to its interaction with growth factors such as basic fibroblast growth factor (bFGF) and transforming growth factor-β (TGF-β). Fucoidan may, therefore, be able to modulate growth factor-dependent pathways in the cell biology of tissue repair (O'Leary et al., 2004). In recent years, the research on drug and gene delivery systems, diagnostic microparticles and wound and burn healing formulations of fucoidan has been increasing in course of time (Sezer et al., 2008a and 2008b, Sezer & Akbuğa, 2009).

Poly-N-acetyl glucosamine

Poly-N-acetyl glucosamine which is produced from marine microalgae, has hemostatic activity and are used as a support material in the treatment of burns and wounds (Pietramaggiori et al., 2008).

b. Synthetic polymers

Polyurethanes and their derivatives

Polyurethanes are copolymers containing urethane groups in their structures. In general, they are formed by conjugation of diol groups and diisocyanate groups with polymerisation reaction (Trumble et al., 2002). A large number of non-toxic polyurethanes are synthesized for use in biomedical applications. One of these, Pellethane 2363-80A which accelerates re-epithelialisation was used as dressing material in the treatment of burn and wound (Wright et al., 1998; Pachulski et al., 2002).

Teflon

Teflon is a polymer that is synthesized by polymerization of tetrafluoroethylene at high temperature and pressure. Teflon is an inert material which is non-carcinogenic, insoluble in polar and nonpolar solvents, and which can be sterilized. It can take the desired shape by application of low-pressure and can easily be applied to the injured area (Raphael et al., 1999; Lee & Worthington, 1999).

Proplast

Proplast which is the first synthetic biomaterial, specially developed for implant applications. It is among the particularly preferred materials in wound, burn and surgical applications due to its high biocompatibility with tissue (Şenyuva et al., 1997).

Methyl methacrylate

Methyl methacrylate is a non-biodegradable synthetic polymer that is resistant to heat and UV. It is used as a dressing and supporting material in plastic surgery and the treatment of injuries (Nakabayashi, 2003).

Silicon

Silicon is used extensively for biomedical purposes. It has low toxicity, low allergic properties and high biocompatibility in the body (O'Donovan et al., 1999; Jansson & Tengvall, 2001). This polymer, which is resistant to biodegradation, is used in the preparation of implant elastomers used in soft tissue repair and in the production of hypodermic needles and syringes (Van den Kerckhove et al., 2001; Park et al., 2002). In addition, silicon is also often used as wound support material in severe wounds and burns due to its high tissue compatibility (Whelan, 2002; Losi et al., 2004).

4.3 Pharmaceutical formulations used as dressings for wounds and burns

Many pharmaceutical formulations have been recently developed as synthetic dressing material for wound and burn treatment.

4.3.1 Films/membranes

These pharmaceutical dosage forms, which are available in thickness ranging from μm to mm, are prepared by different methods using one or more polymers. Films are ideal dressing materials and available in commercial. Films/membranes with a homogeneous polymeric network structure are used to treat the damaged area and generally protect the wound and burn area against external factors (Verma & Iyer, 2000; Stashak et al., 2004). The polymers used in the preparation of films include; polyurethane, polyvinylpyrrolidone (Yoo

& Kim, 2008), hyaluronic acid (Xu et al., 2007; Uppal et al., 2011), collagen (Boa et al., 2008), sodium alginate (Kim et al., 2008a and 2008b) chitosan and its derivatives (Tanigawa et al., 2008), poly-N-acetyl glucosamine (Pietramaggiori et al., 2008) and fucoidan (Sezer et al., 2007). In a clinical study, as a result of *in vitro* studies of polyurethane / poly (N-vinylpyrrolidone) composite film combinations, it was reported that their water absorption capacity was high and water vapour permeability was between 1816-2728 g /m² /day. It was seen that recovery in the injured area was significantly increased and a new epithelial tissue was formed in 15 day period following application of these prepared formulations to full-thickness wounds induced in a rat model (Yoo & Kim, 2008). Tanigawa et al. (2008) prepared scaffold formulations by using chitosan citrate and chitosan acetate, which are natural polymers, and examined the effectiveness of the formulations on wound healing in mice with damaged epidermis. The scaffold, prepared by a lyophilisation method, gave quite a distinct pore structure especially inside it, depending on the type of the acid used for the preparation of chitosan solution; the pore consisted of fibrous networks appeared in chitosan citrate, whereas the pore surround by cell walls occurred in chitosan acetate. Despite the large difference in the pore structure, both scaffolds were effective in regeneration of the outher skin. However, chitosan citrate scaffold provided better facilitation in wound healing than the chitosan acetate one (Tanigawa et al., 2008). Commercially available collagen-based film formulations also used in the treatment of dermal burns (Boa et al., 2008). In a recent study, the treatment efficacy of collagen-based films was investigated in second-degree burns created in 45 Wistar rats. The prepared collagen film formulations were applied to the wound alone or in combination with liposome formulations containing usnic acid, and then examined after 14 and 21 days. The use of the usnic acid provided more rapid substitution of type-III for type-I collagen on the 14th day, and improved the collagenisation density on the 21st day. It was concluded that the use of reconstituted bovine type-I collagen-based films containing usnic acid improved burn healing process (Nunes et al., 2011). Differences in healing of wounds are summarized in histological sections in Figure 2.

In another study, the researchers hypothesized that a poly-N-acetyl glucosamine (pGlcNAc) fibre patch might enhance wound healing in diabetic (db/db) mice. Wounds dressed with pGlcNAc patches for 1 h closed faster than control wounds, reaching 90% closure in 16.6 days, 9 days faster than untreated wounds. Granulation tissue showed higher levels of proliferation and vascularization after 1 h treatment than the 24 h and left-untreated groups. Foreign body reaction to the material was not noted in applications up to 24 h (Pietramaggiori et al., 2008).

The primary negative factor affecting healing of burns and wounds is the loss of skin integrity. It is indicated in the literature that wound and burn healing occur much more quickly with the help of a dressing material (Mutsaers et al., 1997; Kapoor & Appleton, 2005). Therefore, in recent years, the use of biopolymers has gained priority in tissue engineering and biotechnology, both as dressing material and in terms of enhancing treatment efficiency (Atiyeh et al., 2005).

A biopolymer, fucoidan, consisting of fucose and sulphate groups has been used for the treatment of burns and wounds (Sezer et al., 2007, 2008a and 2008b). It was reported that fucoidan shows anticoagulant effect and heparin activity (Sezer & Cevher 2011). It was also reported that films prepared with fucoidan do not prevent contact of the wound surface with air oxygen, provide the moisture balance in wound/burn area, accelerate the migration of fibroblasts and provide re-epithelialisation (Sezer et al., 2007) (Table 5).

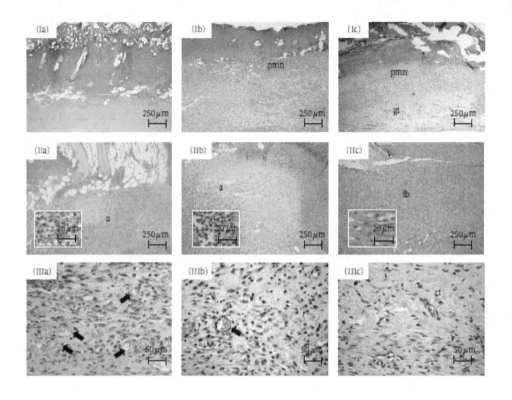

Fig. 2. Histological sections stained in hematoxylen-eosin. Seven days: (Ia) Lack of inflammatory response in the center of the burned area in COL. (Ib) Inflammatory infiltrate rich in polymorphonuclear neutrophils (pmn) in PHO. (Ic) Expressive content of polymorphonuclear neutrophils (pmn) in the top of the wound, and early granulation tissue (gt) formation in the bottom, in UAL. Fourteen days: (IIa and IIb) Intense acute inflammatory reaction (a) scattered within the burned area in COL and PHO, respectively. Neutrophils seen in detail. (IIc) Moderate infiltrate of neutrophils and lymphocytes in association to expressive fibroblastic proliferation (fb) in UAL. Fibroblasts seen in detail. Twentyone days: (IIIa and IIIb) Vascular component (arrows), and chronic inflammatory infiltrate, still evident in COL and PHO, respectively. (IIIc) Scanty inflammatory cells are seen within the cicatricial tissue (ct) in UAL. *COL—animals treated with collagen-based films; PHO—animals treated with collagen films containing empty liposomes; UAL—animals treated with collagen-based films containing usnic acid incorporated into liposomes (Nunes et al., 2011).

	Day 7				Day 14				Day 21			
Codes	Fibroblast	Collagen	MN	PMNL	Fibroblast	Collagen	MN	PMNL	Fibroblast	Collagen	MN	PMNL
Control	+ (6/6)	− (5/6)	− (5/6)	+ (2/6)	+ (4/6)	− (4/6)	− (4/6)	+ (3/6)	+ (4/6)	+ (3/6)	+ (6/6)	+ (5/6)
		+ (1/6)	+ (1/6)	++ (2/6)	++ (2/6)	+ (2/6)	+ (2/6)	++ (2/6)	++ (1/6)	++(3/6)		++ (1/6)
				+++(2/6)				+++ (1/6)	+++1/6			
FS	++ (6/6)	− (1/6)	− (5/6)	− (2/6)	+ (1/6)	+ (3/6)	− (2/6)	− (4/6)	+ (6/6)	+ (4/6)	− (4/6)	− (4/6)
	+ (5/6)	+ (1/6)	+ (1/6)	+ (4/6)	++ (5/6)	++ (3/6)	+ (4/6)	+ (2/6)		++ (2/6)	+ (2/6)	+ (2/6)
CF	+ (6/6)	− (4/6)	− (5/6)	+ (4/6)	+ (4/6)	+ (6/6)	− (4/6)	+ (4/6)	++ (5/6)	++ (4/6)	− (2/6)	− (3/6)
	+ (2/6)	+ (1/6)	++ (2/6)	++ (2/6)		+ (2/6)	++ (2/6)	+++(1/6)	+++(2/6)	+ (3/6)	+ (3/6)
											++ (1/6)	
CFF	+ (6/6)	+ (6/6)	− (3/6)	+ (3/6)	+ (1/6)	++ (4/6)	+ (3/6)	− (6/6)	− (4/6)	− (5/6)	− (6/6)	− (6/6)
			+ (2/6)	++ (3/6)	++ (1/6)	+++(2/6)	++ (2/6)		+ (2/6)	+ (1/6)		
			++(1/6)		+++(4/6)		+++(1/6)					

*MN indicates mononuclear leukocyte; PMNL, polymorphonuclear leukocyte; —, absent; +, mild; ++, moderate; +++, severe; FS, fucoidan solution; CF, chitosan film without fucoidan; and CFF, chitosan film containing fucoidan.

Table 5. The score of the wound cells and collagen (Sezer et al., 2007).

4.3.2 Gels

Gels are viscous semi-solid preparations formed by dispersion of inorganic or organic substances that have larger size than colloidal particles in a liquid phase. Hydrogels are semi-solid systems, formed by a combination of one or more hydrophilic polymer. They are among the dressing materials frequently used in the treatment of wounds and burns. As they are capable of absorbing much more water than their weight, they act as dressing, reducing potential irritation when in contact with tissue and other similar structures. They keep moisture at the application site and permit oxygen penetration (Hoffman, 2002; Jeong et al., 2002). Hydrogels have many advantages including patient compliance, treatment efficacy and ease of application. The advantages of hydrogels in wound and burn treatment can be listed as follows (Kumar et al., 2001; Hoffman, 2002; Jeong et al., 2002; Byrne et al., 2002);

- Bioadhesion of gels to the surface of the wound is high and this also eases the treatment due to increased contact with the wound
- Their structures facilitate the moisture and water vapour permeability necessary to heal the wound area
- Difficulties that are particularly related to the application to open wounds are not seen in these preparations
- They can easily be removed from the application site when adverse events seen

Natural polymers are generally preferred in the preparation of hydrogels. Hyaluronan is a biopolymer widely used in the treatment of wounds. It is non-toxic, non-immunogenic, and has very good resorption characteristics in biomedical applications, which allow this biopolymer to be used in the treatment of wounds. In a study, cross-linked glycol chitosan/hyaluronan hydrogels was prepared and found that they displayed the characteristics required of an ideal wound dressing material (Wang, 2006). Chitosan is obtained by partial deacetylation of the amines of chitin. Its use has been explored in various biomaterial and medical applications. Chitosan has desirable qualities, such as hemostasis, wound healing, bacteriostatic, biocompatibility, and biodegradability properties. Chitosan appears to have no adverse effects after implantation in tissues and, for this reason, it has been used for a wide range of biomedical applications. Chitosan was also used to inhibit fibroplasia in wound healing and to promote tissue growth and differentiation in culture (Alsarra, 2009).

The efficacy of chitosans with different molecular weights and deacetylation degrees was investigated in the treatment of wounds (Sezer, 2011). The treatment efficacy of gels prepared with chitosans with low, medium and high molecular weight was examined in a rat full-thickness wound model in which the epidermis and dermis had been damaged. Chitosan gel formulations were also compared with Fucidin® ointment containing fusidic acid. Chitosan was found to promote the migration of the inflammatory cells which are capable of the production and secretion of a large repertoire of pro-inflammatory products and growth factors at a very early phase of healing. Fucidin® ointment-treated rats revealed a site that was not completely healed but more improved and with a smaller lesion than that of untreated groups. In comparison with high molecular weight chitosan-treated wounds (after 12 days), the wound site was so perfectly healed that it was difficult to distinguish it from normal skin (Figure 3) (Alsarra, 2009).

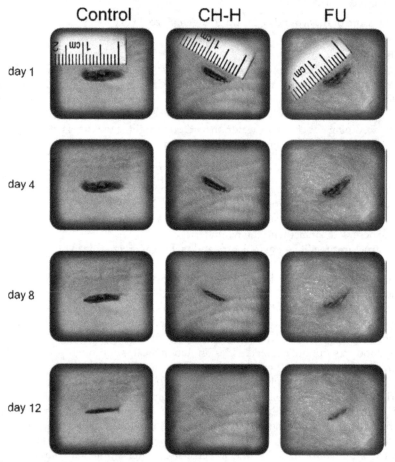

Fig. 3. Photographs of macroscopic appearances of wound excised from rats that were untreated (control), treated with high molecular weight chitosan (CH-H), or treated with Fucidin® ointment (FU) (Alsarra, 2009).

The efficacy of chitosan gel formulations containing silver sulfadiazine in the treatment of burn was examined. Due to its pseudoplastic characteristic and high bioadhesiveness, the chitosan gels with and without silver sulfadiazine showed a satisfactory retention time over the wounds. Wounds treated with chitosan gel with silver sulfadiazine showed a higher fibroblast production and a better angiogenesis than that of commercially available silver sulfadiazine cream, which are important parameters on the evolution of the healing process (Nascimento et al., 2009).

Polyvinyl alcohol hydrogels with different combinations, prepared with chitosan and dextran, are also used as wound dressing material. It was recorded that these cross-linked hydrogels are capable of ideal water absorption and swelling and provide an ideal moist environment necessary for wound healing. In addition, it has been reported in the literature that cross-linking of polyvinyl alcohol with dextran and chitosan increases the flexibility and elasticity of the gel (Sung et al., 2010; Hwang et al., 2010).

In another study, the novel thermoreversible wound gel formulation containing Polyhexadine showed good results for wound treatment and represented an alternative to silver sulfadiazine cream and iodine-based ointment, which are toxic and associated with many allergic reactions. The open randomized controlled single-center study was carried out on 44 patients in 2 parallel groups (Traumasept®wound covering gel vs. Flammazine®). The transparency of the thermosensitive gel formulation and absence of staining allowed for good wound assessment, without the need for painful cleaning, and the change from the fluid to the gel state enabled easy handling and filling the wound cavities (Goertz et al., 2010).

One of the most important parameters of the injured tissue pathology is to stop the bleeding and to protect the area with a protective support material in the first step. It is reported that hydrogels prepared using oxidised dextran and allylamine hydrochloride decreased coagulation and increased cloth strength and consequently are able to decrease the haemorrhage in clinical use (Peng & Shek, 2009).

The use of various proteins in topical wound treatment is also recorded in the literature (Ji et al., 2009). Wound treatment efficacy of methylcellulose gel dressings including recombinant human vascular endothelial growth factor (rhVEGF) have been examined by Ji et al. (2009). The dressings being studied were Adaptic®, Non-stick Dressing, Conformant 2®, Opsite®™ and Tegapore™. The criteria to select a compatible dressing include protein stability, absence of leachables from the dressing, and ability to retain gel on wound. Results showed that rhVEGF was significantly oxidized by Adaptic dressing in 24 h. Protein oxidation was likely due to the peroxides, as determined by ferrous oxidation with xylenol orange (FOX) assay, released into the protein solution from the dressing. In conclusion, Tegapore™ was considered suitable for the rhVEGF topical gel (Ji et al., 2009).

A similar study determined the gel properties of sodium carboxymethyl cellulose and fusidic acid and aimed to improve them as wound dressings. It was reported that swelling, flexibility and elasticity properties of sodium carboxymethyl cellulose (Na-CMC) gels was changed depending on sodium fucidate and the cross linker (PVA) content in the formulation. Hydrogel formulation containing 2.5% PVA, 1.125% Na-CMC and 0.2% sodium fucidate adsorbed exudate from the wound surface adequately and kept the moisture sufficiently in wound area and consequently, it was recommended as an ideal wound dressing material. (Lim et al., 2010).

Hydroxyapatite (HA) and silk fibroin (SF) composites are biomaterials used in wound treatment. It was reported that the polarized HA (pHA) transforms the SF structure into a porous three-dimensional scaffold. SF gel containing pHA was found to be higher promotive effects on wound healing, re-epithelialisation and matrix formation (Okabayashi et al., 2009).

Gelatine is commonly used in the treatment of wounds and burns, and acts as supportive tissue when used in the treatment of dermal burns with various biopolymers. Balakrishnan et al. (2005) applied hydrogel formulations containing oxidised alginate and gelatine to the 1 cm² full-thickness skin wounds created in rat model and the results evaluated histologically. At 15th day, test wounds appeared reduced in size with new epithelium noted at both the edges of the defect with the proliferation of basal layer and formation of the rete pegs. New collagen formed in the dermis appeared mature. Granulation tissue was seen in dermis. Granulation tissue formation is essential for permanent wound closure, since it fills the defects and prepares the way for re-epithelialisation. These findings support that alginate/gelatin hydrogel is able to provide suitable condition for granulation tissue formation. At 15th day, in test wounds, the defect area became smaller and filled with fibro-proliferative tissue. Inflammatory cells were absent. However, for some control wounds though the entire surface of the defect was covered with new epithelium, moderate number of inflammatory cells, predominantly lymphocytes and macrophages, were still present in the upper dermis. Though superficially neither control nor test wounds showed any reduction in defect area at 5th day, on measuring the wound re-epithelialisation it was found that both wounds have started healing (Balakrishnan et al., 2005).

In a previous study has shown successful treatment with fucoidan-chitosan hydrogels which were tested in New Zealand rabbits with second degree burn (Sezer et al., 2008a). In another study, chitin/chitosan, fucoidan and alginate hydrogel blends were prepared and the granulation tissue and capillary formation were found to be increased in the first 7 days of the treatment of induced wounds (Murakami et al., 2010).

4.3.3 Sprays and foams

Sprays are pharmaceutical forms containing the solvent and polymer, forming a film layer on the surface of the wound when sprayed. The best example of a spray-based artificial wound and burn dressing is Hydron. It is prepared with polyhydroxyethyl methacrylate powder and liquid polyethylene glycol. When it is sprayed on the surface of the wound, it creates a thin and transparent film layer. In studies it was found that sprays reduce the pain of the wound, but have disadvantages including loss of integrity of the dressing and accumulation of sub-membrane fluid. Researchers stated that Hydron provides an effective treatment when applied to small partial thickness wounds and to areas which are away from joints (Dressler et al., 1980; Pruitt & Levine, 1984). Another example of aerosol sprays is papain-pectin sprays. The spray-on topical wound debrider composition consisting of 0.1% papain immobilized in 6% pectin gel was formulated for skin wound healing. The stability of the enzyme activity of this new formulated spray was compared with the 0.1% papain in water solution at the refrigerated temperature of 4C. and 75°C. Prepared formulations were tested on experimental wounds created on rabbits. In the study groups treated by pectin-papain aerosol spray compared with a control group.

During the experiment, no obvious healing process inhibition or side effects were visually observed. Upon spray application on the surgical wound, the aerosol formed a thin, smooth, and even film staying in place on the wound bedwithout dripping, promoting wound healing after drying versus an untreated wound (as control). The spray bottle was easy to maneuver; making it possible to reach areas of each wound that otherwise might receive inadequate coverage. The progress of healing was overall higher with the spray at 2 times more in the first four days of treatment. The difference was calculated to be significant based on the Student´s t-test method with the resulting $p < 0.05$. It was concluded that papain immobilized in pectin can be used for the development of effective aerosol spray system for wound healing in the areas of enzymatic debridement of necrotic tissue and liquefaction of slough to remove dead or contaminated tissue in acute and chronic lesions, such as diabetic ulcers, pressure ulcers, varicose ulcers, and traumatic infected wounds, postoperative wounds, burns, carbuncles, and pilonidal cyst wounds (Jáuregui et al., 2009).

Lyofoam, polyurethane foam, is normally hydrophobic; however, when applying heat and pressure, it becomes hydrophilic and, in this form, while providing blood and exudates absorption, it also prevents drying the wound surface completely (Johnson et al., 1998; Catarino et al., 2000; Fenn & Butler, 2001; Lehnert & Jhala, 2005).

4.3.4 Composites

Composites developed for wound treatment may involve an elastic outer layer with high mechanical strength, which is resistant to the effects of the environment and provides moisture by preventing evaporation; in contrast, the inner layer provide adhesion of the composite to the surface of the wound. Telfa™ is a dressing material, including cotton, covered by polyester film and is used both for providing absorption and preventing dehydration of the wound surface (Kickhöfen et al., 1986). Clinical studies have been conducted of chitin nanofibrils/chitosan glycolate composites (Muzzarelli et al., 2007), salmon milt DNA/salmon collagen composites (Shen et al., 2008), polymer-xerogel composites (Costache et al., 2010), and autologous cellular gel matrix systems (Weinstein-Oppenheimer et al., 2010). Chitin and chitosan composites were found to be very promising in the treatment of wounds. Muzzarelli et al. (2007) tested the wound treatment activity of spray, gel and gauze forms of nanofibrils chitin/chitosan glycolate composites on both Wistar male rats and 75 patients between the ages of 45-70. Recovery was particularly good when applied gauze to gangrenous tissue (Figure 4).

It was shown that the nanofibrillar chitin/chitosan glycolate composites appeared to be most suitable as medicaments able to exert control over various biochemical and physiological processes involved in wound healing besides haemostasis. Whereas chitosan provided antimicrobial activity, cell stimulation capacity and filmogenicity, chitin nanofibrils restructured the gel, released N-acetylglucosamine slowly and recognised proteins and growth factors.

In another study, Shen et al. (2008) examined the neovascularization and fibrillogenesis effects of salmon milt DNA and salmon collagen (SC) composites when used for the treatment of wounds. Tissue loss of wounds treated with composites was repaired quickly and the epidermal layer was formed quickly by means of sDNA (Shen et al., 2008).

Biobrane, is a collagen-silicone based composite, uses as a skin graft to treat injuries (Figure 5). The outer layer of the membrane is a thin and semi-permeable layer consisting of silicon. This layer allows water permeation but prevents the entry of microorganisms. Type I pig collagen forms an inner layer with an inert, hydrophilic network structure and provides a suitable platform for the development of granulation tissue. Water transfer can be maintained similar to that of natural skin by modifying the membrane thickness (Suzuki et al., 1990; Ou et al., 1998; Still et al., 2003).

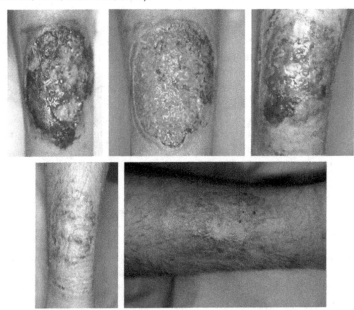

Fig. 4. Gangrenous pyoderma on tibial surface. Treated with Gauze and i.v. therapy of steroids and cyclosporine to ameliorate the wound bed. Complete healing in 40 days (Muzzarelli et al., 2007).

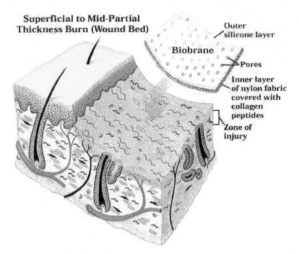

Fig. 5. Schematic observation of biobrane application on superficial to mid-partial thickness burn (http://www.burnsurgery.com).

Bilaminar composite membranes containing bovine collagen-based dermal analogue and silastic epidermis has been developed for the treatment of wounds and burns. It is stated that biocompatible bovine collagen-based dermal analogue slowly degrades and provides a suitable environment for the development of the patient's connective tissue; in addition, the epidermal layer's water vapour permeability is close to the skin and protects the wound from trauma and microorganisms (Still et al., 2003).

There are certain restrictions in the areas in which skin source materials can be applied as dressings. Biodegradable or non-biodegradable unilaminar and bilaminar composite membranes function as a serous crust and reduce pain but can not prevent infection. It is stated that synthetic composites do not constitute infection, are suitable for use with antimicrobial solutions and are easily applied to wound tissue (Van den Kerckhove et al., 2001; Van Zuijlen et al., 2003).

4.3.5 Particulate systems

The biggest advantages of particulate systems are that, when applied locally to open wounds, they easily provide water vapour and oxygen permeability of the wound; and have large contact surfaces and high bioadhesiveness due to their multiparticulate structures. Drug release in the wound area can be controlled with particulate system, and this increases the speed of wound healing (Kawaguchi, 2000; Date & Patravale, 2004).

In recent studies on micro-/nanoparticulate systems in wound and burn treatment, the use of nitric oxide nanoparticles (Martinez et al., 2009), poly (ethylene-co-vinyl alcohol) nanofiber (Xu et al., 2011), silver nanoparticles (Lakshmana et al., 2010; Xu et al., 2011), fucoidan microparticles (Sezer et al., 2008b), collagen sponges (Still et al., 2003; Lee, 2005), and liposomes containing epidermal growth factor (Alemdaroglu et al., 2008) have examined. *Staphylococcus aureus* is a gram-positive bacteria, capable of rapidly proliferating in the injured area and causing infection, causes superficial and invasive skin infections. Wounded skin is suitable media for the growth of such pathogenic microorganisms. Various clinical studies have attempted to develop strategies and formulations to address this common issue of pathogenic infection. Topically applied nitric oxide (NO) is a potentially useful preventive and therapeutic strategy against superficial skin infections, including methicillin-resistant *Staphylococcus aureus* infections. NO modulates immune responses and is a significant regulator of wound healing (Martinez et al., 2009). NO nanoparticles were prepared by combination of sodium nitrite with tetramethylorthosilicate, polyethylene glycol, chitosan, glucose. The mechanisms through which the NO nanoparticles accelerate wound healing were further determined by establishing whether NO nanoparticles prevented collagen degradation by MRSA in the infected tissue. Collagen content was highest in both uninfected and infected wounds treated with NO nanoparticles, although nanoparticles-treated uninfected tissue also had high collagen content. The dispersed blue stain indicated thicker and more mature tissue collagen formation in wounds treated with NO nanoparticles, suggesting that NO nanoparticles exposure maintained dermal architecture through bacterial clearance, and ultimately by guarding collagen (Figure 6) (Martinez et al., 2009).

Silver has been used in wound treatment since ancient times. Ointments including silver sulfadiazine are also frequently used in the treatment of burns. Silver affects pathogenic bacteria in wound and burn areas in different ways. Silver ions interacting with bacterial enzymes are taken up inside the bacterial cells, impair the DNA of the bacteria and prevent cell proliferation. Silver ions also attach the cell wall and disrupt the integrity of cell membrane and kill the bacteria (Klasen, 2000a and 2000b). Poly (ethylene-co-vinyl alcohol) fibre systems including silver nanoparticles were prepared for the treatment of wounds. The results showed that the nanofibre size can be controlled by regulating polymer solution concentration. It has been reported that high concentration of silver might change the fibre morphology. Results of bacterial tests showed that pathogen-restraining ability of the silver-encapsulated nanofibres was effective and proportional over a range of silver concentration, indicating its inflammation control capacity and the potential for applications in skin wound treatment (Xu et al., 2011).

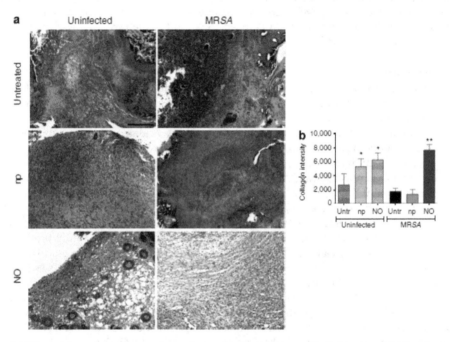

Fig. 6. NO-nps decrease collagen degradation in skin lesions of Balb/c mice. (a) Histological analysis of Balb/c mice uninfected and untreated, uninfected treated with nanoparticles without nitric oxide (NO) (np), uninfected treated with NO-nps (NO), untreated methicillin-resistant Staphylococcus aureus (MRSA)-infected, np-treated MRSA-infected, and MRSA-infected treated with NO, day 7. Mice were infected with 107 bacterial cells. The blue stain indicates collagen. Bar ¼ 25 mm. (b) Quantitative measurement of collagen intensity in 16 representative fields of the same size for uninfected and untreated, uninfected treated with nanoparticles without NO (np), uninfected treated with NO, untreated MRSA-infected, np-treated MRSA-infected, and MRSA-infected treated with NO wounds. Bars are the averages of the results, and error bars denote SDs. *P<0.01 in comparing the untreated groups with the uninfected np- and NO-treated groups; **P<0.001 in comparing the untreated groups with the MRSA + NO group (Martinez et al., 2009).

In another study, silver nanoparticles were synthesized by aqueous and organic methods and incorporated into electrospun polyurethane (PU) nanofibre to enhance the antibacterial as well as wound healing properties. The electrospinning parameters were optimized for PU with and without silver nanoparticles. The water absorption, antibacterial and cytocompatibility of the PU-silver nanofibers were studied and compared to that of conventional PU foam. The results indicated that the PU-Ag nanofibers could be used for wound dressing applications (Lakshmana et al., 2010).

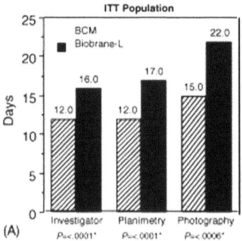

(A)

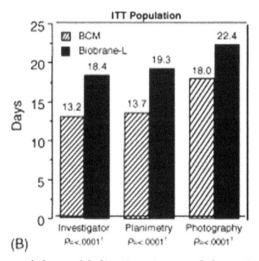

(B)

Fig. 7. Days to 100% wound closure. Median times to wound closure is shown in (A) and mean times are shown in (B). BCM indicates bilayered cellular matrix (OrCelTM); intent-to-treat (ITT). *Log-rank test of the difference between treatment healing times, stratified by patient. †Paired t-test (Still et al., 2003).

In different stages of wound healing, various cell types, cytokines, coagulation factors, growth factors, complement activation and matrix proteins are involved in different extents. Collagen is one of the most important structural protein components of connective tissue. It is difficult to say at which stage collagen predominates because wound healing is a dynamic event involving many stages. There are many different clinical applications of collagen. For example, collagen gel is often used to treat haemophilia patients due to its haemostatic effects (Tan et al., 2001; Lee et al., 2002; Ruszczak, 2003; Beckenstein et al., 2004). By using microsponge technology, which is a new type of pharmaceutical dosage form, collagen-based particles are clinically tested in the treatment of wounds and burns and were reported to have positive results compared with commercial preparations (Ruszczak, 2003; Beckenstein et al., 2004). Still et al. (2003) examined the safety and efficacy of porous collagen sponge containing co-cultured allogeneic donor epidermal keratinocytes and dermal fibroblasts from human neonatal foreskin tissue (OrCel™) in facilitating timely wound closure of split-thickness donor sites in severely burned patients. Utilized a matched pairs design; each patient had two designated donor sites of equivalent surface area and depth. Sites were randomized to receive a single treatment of either OrCel™ or the standard dressing Biobrane-L®. The treatment of donor site wounds with OrCel™ is well tolerated, promotes more rapid healing, and results in reduced scarring when compared with conventional therapy with Biobrane-L® (Still et al., 2003). Data on wound closure as a result of the treatment are given in Figure 7. As seen in Figure 7, the cell based-membrane sponges improved the healing of the wound.

The efficacy of microsponges containing epidermal growth factor (EGF) was investigated clinically. It was found that the application of EGF microsponges to the burn surface increased the activity of fibrinogen and thus fibroblast synthesis and migration were seen at the injury site (Lee, 2005). Also, liposome systems in which EGF had been encapsulated was used as particulate carrier systems in the treatment of wounds and burns. Positive results were reported in terms of eschar tissue formation and wound healing with these systems, when especially used in the treatment of second-degree burns (Alemdaroglu et al., 2008).

5. Conclusion

The biopolymers are more effective as a wound-healing accelerator than synthetic polymers. The wound treated with biopolymers and biomaterials shows accelerated healing. Biopolymers structural arrangement is similar to that of normal skin. Consequently, the biopolymers are considered to be one of ideal materials with biocompatibility, biodegradability, and wound healing property as well as easy application.

6. References

Alemdaroglu, C.; Degim, Z.; Çelebi, N.; Şengezer, M.; Alömeroglu, M. & Nacar, A.(2008). Investigation of epidermal growth factor containing liposome formulation effects on burn wound healing, *Journal of Biomedical Materials Research*, Vol.85A, pp.271–283.

Alper, J.C.; Welch, E.A.; Ginsberg, M.; Bogaars, H. & Maguire, P. (1983). Moist wound healing under a vapor permeable membrane, *Journal of the American Academy of Dermatology*, Vol.8, pp.347-353.

Alsarra, I.A. (2009). Chitosan topical gel formulation in the management of burn wounds, *International Journal of Biological Meacromolecules*, Vol.45, pp.16-21.

Atiyeh, B.S.; Hayek, S.N. & Gunn, S.W. (2005). New technologies for burn wound closure and healing-review of the literature, *Burns*, Vol.31, pp.944-956.

Balasubramani, M.; Kumar, T.R. & Babu, M. (2001). Skin substitutes: a review. *Burns*, Vol.27, pp.534-544.

Balakrishnan, B.; Mohanty, M.; Umashankar, P.R. & Jayakrishnan, A. (2005). Evaluation of an in situ forming hydrogel wound dressing based on oxidized alginate and gelatin, *Biomaterials*, Vol.26, No.32, pp.6335-6342.

Bao, L.; Yang, W.; Mao, X.; Mou, S. & Tang, S. (2008). Agar/collagen membrane as skin dressing for wounds, *Biomedical Materials*, Vol.3, pp.1-7.

Barnett, A; Berkowitz, R.L.; Mills, R. & Vistnes, L.M. (1983). Scalp as skin graft donor site: rapid reuse with synthetic adhesive moisture vapour permeable dressings, *Journal of Trauma*, Vol.23, pp.148-151.

Beckenstein, M.S.; Kuniaki, T. & Matarasso, A. (2004). The effect of scarguard on collagenase levels using a full-thickness epidermal model, *Aesthetic Surgery Journal*, Vol.24, pp.542-546.

Berthod, F.; Saintigny, G.; Chretien, F.; Hayek, D.; Collombel, C. & Damour, O. (1994). Optimization of thickness, pore size and mechanical properties of a biomaterial designed for deep burn coverage, *Clinical Materials*, Vol.15, pp.259-265.

Branski, L.K.; Herndon, D.N.; Celis, M.M.; Norbury, W.B.; Masters, O.E. & Jeschke, M.G. (2008). Amnion in the treatment of pediatric partial-thickness facial burns, *Burns*, Vol.34, No.3, pp.393-399.

Byrne, M.E.; Park, K. & Peppas, N.A. (2002). Molecular imprinting within hydrogels, *Advanced Drug Delivery Reviews*, Vol.54, pp.149-161.

Catarino, P.A.; Chamberlain, M.H.; Wright, N.C.; Black, E.; Campbell, K.; Robson, D. & Pillai, R.G. (2000). High-pressure suction drainage via a polyurethane foam in the management of poststernotomy mediastinitis, *Annals of Thoracic Surgery*, Vol.70, pp.1891-1895.

Chanson, M.; Derouette, J.; Roth, I.; Foglia, B.; Scerri, I.; Dudez, T. & Kwak, B.R. (2005). Gap junctional communication in tissue inflammation and repair, *BBA*, Vol.1711, pp.197-207.

Cho, Y.W.; Cho, Y.N.; Chung, S.H.; Yoo, G. & Ko, S.W. (1999). Water-soluble chitin as a wound healing accelerator, *Biomaterials*, Vol.20, pp.2139-2145.

Chvapil, M.; Kronenthal, R,L. & Van Winkle, W. (1973). Medical and surgical applications of collagen, *International Review of Connective Tissue Research*, Vol.6, pp.1–61.

Costache, M.C.; Qu, H.; Ducheyne, P. & Devore, D.I. (2010). Polymer-xerogel composites for controlled release wound dressings, *Biomaterials*, Vol.31, No.24, pp.6336-6343.

Date, A.A. & Patravale, V.B. (2004). Current strategies for engineering drug nanoparticles, *Current Opinion in Colloid and Interface Science*, Vol.9, pp.222-235.

Dressler, D.P.; Barbee, W.K. & Sprenger, R. (1980). The effect of Hydron® burn wound dressing on burned rat and rabbit ear wound healing, *Journal of Trauma*, Vol.20, pp.1024-1028.

Fansler, R.F.; Taheri, P.; Cullinane, C.; Sabates, B. & Flint, L.M. (1995). Polypropylene mesh closure of the complicated abdominal wound, *American Journal of Surgery*, Vol.170, pp.15-18.

Fenn, C.H. & Butler, P.E.M. (2001). Abdominoplasty wound-healing complications: assisted closure using foam suction dressing, *British Journal of Plastic Surgery*, Vol.54, pp.348-351.

Freyman, T.M.; Yannas, I.V. & Gibson, L.J. (2001). Cellular materials as porous scaffolds for tissue engineering, *Progress in Materials Science*, Vol.46, pp.273-282.

Gangjee, T.; Colaizzo, R. & Von Recum, A.F. (1985). Species-related differences in percutaneous wound healing, *Annals of Biomedical Engineering*, Vol.13, No.5, pp.451-467.

Gingras, M.; Paradis, I. & Berthod, F. (2003). Nerve regeneration in a collagen-chitosan tissue-engineered skin transplanted on nude mice, *Biomaterials*, Vol.24, pp.1653-1661.

Goertz, O.; Abels, C.; Knie, U.; May, T.; Hirsch, T.; Daigeler, A.; Steinau, H.U. & Langer, S. (2010). Clinical safety and efficacy of a novel thermoreversible polyhexanide-preserved wound covering gel, *European Surgical Research*, Vol. 44, No.2, pp.96-101.

Hafemann, B.; Ensslen, S.; Erdmann, C.; Niedballa, R.; Zühlke, A.; Ghofrani, K. & Kirkpatrick, C.J. (1999). Use of a collagen/elastin-membrane for the tissue engineering of dermis, *Burns*, Vol.25, pp.373-384.

Hasegawa, T.; Mizoguchi, M.; Haruna, K.; Mizuno, Y.; Muramatsu, S.; Suga, Y.; Ogawa, H. & Ikeda, S. (2007). Amnia for intractable skin ulcers with recessive dystrophic epidermolysis bullosa: report of three cases, *Journal of Dermatology*, Vol. 34, pp.328–332.

Hoffman, A.S. (2002). Hydrogels for biomedical applications, *Advanced of Drug Delivery Reviews*, Vol.43, pp.3-12.

http://www.burnsurgery.com

Hwang, M.R.; Kim, J.O.; Lee, J.H.; Kim, Y.I.; Kim, J.H.; Chang, S.W.; Jin, S.G.; Kim, J.A.; Lyoo, W.S.; Han, S.S.; Ku, S.K.; Yong, C.S. & Choi, H.G. (2010). Gentamicin-loaded wound dressing with polyvinyl alcohol/dextran hydrogel: gel characterization and *in vivo* healing evaluation, *AAPS PharmSciTech*, Vol.11, No.3, pp.1092-1103.

Integument. (2001). Ed, L.P. Gartner & J.L. Hiatt, Color Textbook of Histology. Vol. 14, pp. 325-342, W.B. Saunders Company, Philadelphia, USA.

Jansson, E. & Tengvall, P. (2001). *In vitro* preparation and ellipsometric characterization of thin blood plasma clot films on silicon, *Biomaterials*, Vol.22, pp.1803-1808.

Jáuregui, K.M.G.; Cabrera, J.C.C.; Ceniceros, E.P.S.; Hernández, J.L.M. & Ilyina, A. (2009). A new formulated stable papain-pectin aerosol spray for skin wound healing, *Biotechnology and Bioprocess Engineering*, Vol.14, pp.450-456.

Jeong, B.; Kim, S.W. & Bae, Y.H. (2002). Thermosensitive sol-gel reversible hydrogels, *Advanced of Drug Delivery Reviews*, Vol.54, pp.37-51.

Ji, J.A.; Borisov, O.; Ingham, E.; Ling, V. & Wang, Y.J. (2009). Compatibility of a protein topical gel with wound dressings, *Journal of Pharmaceutical Sciences*, Vol.98, No.2, pp.595-605.

Johnson, P.A.; Fleming, K. & Avery, C.M.E. (1998). Latex foam and staple fixation of skin grafts, *British Journal of Oral and Maxillofacial Surgery*, Vol. 36, pp.141-142.

Jones, I.; Currie, L. & Martin, R. (2002). A guide to biological skin substitutes, *British Journal of Plastic Surgery*, Vol.55, pp.185-193.

Kapoor, M. & Appleton, I. (2005). Wound healing: abnormalities and future therapeutic targets, *Current Anaesthesia & Critical Care*, Vol.16, pp.88-93.

Kawaguchi, H. (2000). Functional polymer microspheres. *Progress in Polymer Science*, Vol.25, pp.1171-1210.

Kearney J.N. (2001). Clinical evaluation of skin substitutes, *Burns*, Vol.27, pp.545-551.

Khor, E.; Lim, L.Y. (2003). Implantable applications of chitin and chitosan, *Biomaterials*, Vol.24, pp. 2339-2349.

Kickhöfen, B.; Wokalek, H.; Scheel D. & Ruh, H. (1986). Chemical and physical properties of a hydrogel wound dressing, *Biomaterials*, Vol.7, pp.67-72.

Kim, I.; Park, J. W.; Kwon, I.C.; Baik, B.S. & Cho, B.C. (2002). Role of BMP βig-h3, and chitosan in early bony consolidation in distraction osteogenesis in a dog model, *Plastic and Reconstructive Surgery*, Vol.109, pp.1966-1977.

Kim, J.O.; Choi, J.Y.; Park, J.K.; Kim, J.H.; Jin, S.G.; Chang, S.W.; Li, D.X.; Hwang, M.R.; Woo, J.S.; Kim, J.A.; Lyoo, W.S.; Yong, C.S. & Choi, H.G. (2008). Development of clindamycin-loaded wound dressing with polyvinyl alcohol and sodium alginate, *Biological and Pharmaceutical Bulletin*, Vol. 31, No.12, pp.2277-2282.

Kim, J.O.; Park, J.K.; Kim, J.H.; Jin, S.G.; Yong, C.S.; Li, D.X.; Choi, J.Y.; Woo, J.S.; Yoo, B.K.; Lyoo, W.S.; Kim, J. & Choi, H. (2008). Development of polyvinyl alcohol–sodium alginate gel-matrix-based wound dressing system containing nitrofurazone, *International Journal of Pharmaceutics*, Vol.359, pp.79–86.

Kirker, K.R.; Luo, Y.; Nielson, J.H.; Shelby, J. & Prestwich, G.D. (2002). Glycosaminoglycan hydrogel films as bio-interactive dressings for wound healing, *Biomaterials*, Vol.23, pp.3661-3671.

Klasen, H.J. (2000a). A historical review of the use of silver in the treatment of burns. I. early uses, *Burns*, Vol.26, pp.117-130.

Klasen, H.J. (2000b). A historical review of the use of silver in the treatment of burns. II. renewed interest for silver, *Burns*, Vol.26, pp.131-138.

Knapp, T.R.; Kaplan, E.N. & Daniels, J.R. (1977). Injectable collagen for soft tissue augmentation, *Plastic and Reconstructive Surgery*, Vol.60, pp.398-405.

Krajewska, B. (2004). Application of chitin and chitosan based materials for enzyme immobilizations: a review, *Enzyme and Microbial Technology*, Vol.35, pp.126-139.

Kumar, N.; Ravikumar, M.N.V. & Domb, A.J. (2001). Biodegradable block copolymers, *Advanced of Drug Delivery Reviews*, Vol.53, pp.23-44.

Lakshmana, L.R.; Shalumona, K.T.; Naira, S.V.; Jayakumara, R. & Nair, S.V. (2010). Preparation of silver nanoparticles incorporated electrospun polyurethane nano-fibrous mat for wound dressing, *Journal of Macromolecular Science, Part A: Pure and Applied Chemistry*, Vol.47, pp.1012–1018.

Lee, A.R. (2005). Enhancing dermal matrix regeneration and biomechanical properties of 2nd degree-burn wounds by EGF-impregnated collagen sponge dressing, *Archives of Pharmacal Research*, Vol.28, No.11, pp.1311-1316.

Lee, C.H.; Singla, A. & Lee, Y. (2001). Biomedical application of collagen, *International Journal of Pharmaceutics*, Vol.221, pp.1-22.

Lee, J.J. & Worthington, P. (1999). Reconstruction of the temporomandibular joint using calvarial bone after a failed teflon-proplast implant, *Journal of Oral and Maxillofacial Surgery*, Vol.57, pp.457-461.

Lehnert, B. & Jhala, G. (2005). The use of foam as a postoperative compression dressing, *Journal of Foot and Ankle Surgery*, Vol.44, pp.68-69.

Lim, J.K.; Saliba, L.; Smith, M.J.; McTavish, J.; Raine, C. & Curtin, P. (2000). Normal saline wound dressing-is it really normal?, *British Journal of Plastic Surgery*, Vol.53, pp.42-45.

Lim, S.J.; Lee, J.H.; Piao, M.G.; Lee, M.K.; Oh, D.H.; Hwang, du H.; Quan, Q.Z.; Yong, C.S. & Choi, H.G. (2010). Effect of sodium carboxymethylcellulose and fucidic acid on the gel characterization of polyvinylalcohol-based wound dressing, *Archives of Pharmacal Research*, Vol.33, No.7, pp.1073-1081.

Lloyd, L.L.; Kennedy, J.F.; Methacanon, P.; Paterson, M. & Knill, C.J. (1998). Carbohydrate polymers as wound management aids, *Carbohydrate Polymers*, Vol. 37, pp.315-322.

Losi, P.; Lombardi, S.; Briganti, E. & Soldani, G. (2004). Luminal surface microgeometry affects platelet adhesion in small-diameter synthetic grafts, *Biomaterials*, Vol.25, pp.4447-4455.

Madri J.A. (1990). Inflammation and healing. Ed: J. M. Kissane, Anderson's Pathology. Vol. 1, pp. 67-110, The C.V. Mosby Company, St. Louis, USA.

Manwaring, M.E.; Walsh, J.F. & Tresco P.A. (2004). Contact quidance induced organization of extracellular matrix, *Biomaterials*, Vol.25, pp.3631-3638.

Martinez, L.R.; Han, G.; Chacko, M.; Mihu, M.R.; Jacobson, M.; Gialanella, P.; Friedman, A.J.; Nosanchuk, J.D. & Friedman J.M. (2009). Antimicrobial and healing efficacy of sustained release nitric oxide nanoparticles against *Staphylococcus aureus* skin infection, *Journal of Investigative Dermatology*, Vol. 129, No. 10, pp.2463-2469.

Mi, F.; Wu, Y.; Shyu, S.; Schoung, J.; Huang, Y.; Tsai, Y. & Hao J. (2002). Control of wound infections using a bilayer chitosan wound dressing with sustaniable antibiotic delivery, *Journal of Biomedical Materials Research*, Vol.59, pp.438-449.

Miller, R.S.; Steward, D.L.; Tami, T.A.; Sillars, M.J.; Seiden, A.M.; Shete, M.; Paskowski, C. & Welge, J. (2003). The clinical effects of hyaluronic acid ester nasal dressing (Merogel) on intranasal wound healing after functional endoscopic sinus surgery, *Otolaryngology-Head and Neck Surgery*, Vol.128, pp.862-869.

Moulin, V.; Auger, F.A.; Garrel, D. & Germain, L. (2000). Role of wound healing myofibroblasts on re-epithelialization of human skin, *Burns*, Vol.26, pp.3-12.

Mutsaers, S.E.; Bishop, J.E.; McGrouther, G. & Laurent, G.J. (1997). Mechanisms of tissue repair: from wound healing to fibrosis, *International Journal of Biochemistry & Cell Biology*, Vol.29, pp.5-17.

Murakami, K.; Aoki, H.; Nakamura, S.; Nakamura, S.; Takikawa, M.; Hanzawa, M.; Kishimoto, S.; Hattori, H.; Tanaka, Y.; Kiyosawa,, T.; Sato, Y. & Ishihara, M. (2010). Hydrogel blends of chitin/chitosan, fucoidan and alginate as healing-impaired wound dressings, *Biomaterials*, Vol.31, No.1, pp.83-90.

Muzzarelli, C. & Muzzarelli, R.A.A. (2002). Natural and artificial chitosan-inorganic composites, *Journal of Inorganic Biochemistry*, Vol.92, pp.89-94.

Muzzarelli, R.A.A.; Morganti, P.; Morganti, G.; Palombo, P.; Palombo, M.; Biagini, G.; Belmonte, M.M.; Giantomassi, F.; Orlandi, F. & Muzzarelli, C. (2007). Chitin nanofibrils/chitosan glycolate composites as wound medicaments, *Carbohydrate Polymers*, Vol. 70, pp. 274–284.

Naimer, S.A. & Chemla, F. (2000). Elastic adhesive dressing treatment of bleeding wounds in trauma victims, *American Journal of Emergency Medicine*, Vol.18, pp.816-819.

Nakabayashi, N. (2003). Dental biomaterials and the healing of dental tissue, *Biomaterials*, Vol.24, pp.2437-2439.

Nanchahal, J.; Dover, R. & Otto, W.R. (2002). Allogeneic skin substitutes applied to burns patients, *Burns*, Vol.28, pp.254-257.

Nascimento, E.G.; Sampaio, T.B.; Medeiros, A.C. & Azevedo, E.P. (2009). Evaluation of chitosan gel with 1% silver sulfadiazine as an alternative for burn wound treatment in rats, *Acta Cirurgica Brasileira*, Vol.24, No.6, pp.460-465.

Nunes, P.S.; Albuquerque-Júnior, R.L.C.; Cavalcante, D.R.R.; Dantas, M.D.M.; Cardoso, J.C.; Bezerra, M.S.; Souza, J.C.C.; Serafini, M.R.; Quitans-Jr, L.J.; Bonjardim, L.R. & Araújo A.A.S. (2011). Collagen-based films containing liposome-loaded usnic acid as dressing for dermal burn healing, *Journal of Biomedicine and Biotechnology*, Article ID. 761593.

O'Donovan, D.A.; Mehdi, S.Y. & Eadie, P.A. (1999). The role of mepitel silicone net dressings in the managment of fingertip injuries in children, *Journal of Hand Surgery*, Vol.24, pp.727-730.

Okabayashi, R.; Nakamura, M.; Okabayashi, T.; Tanaka, Y.; Nagai, A. & Yamashita, K. (2009). Efficacy of polarized hydroxyapatite and silk fibroin composite dressing gel on epidermal recovery from full-thickness skin wounds, *Journal of Biomedical Materials Research Part B: Applied Biomaterials*, Vol.90B, pp.641–646.

O'Leary, R.; Rerek, M. & Wood, E.J. (2004). Fucoidan modulates the effect of transforming growth factor (TGF)-beta1 on fibroblast proliferation and wound repopulation in in vitro models of dermal wound repair, *Biological and Pharmaceutical Bulletin*, Vol.27, No.2, pp.266-270.

Ortega, M.R. & Milner, S.M. (2000). Human beta defensin is absent in burn blister fluid, *Burn*, Vol.26, pp.724-726.

Ou, L.F.; Lee, S.Y.; Chen, Y.C.; Yang, R.S. & Tang, Y.W. (1998). Use of biobrane in pediatric scald burns-experience in 106 children, *Burns*, Vol.24, pp.49-53.

Pachulski, R.; Zasadil, M.; Adkins, D. & Hanif, B. (2002). High incidence of Pellethane 90A lead malfunction, *Europace*, Vol.4, pp.45-47.

Park, S.; Lee, K.C.; Song, B.S.; Cho, J.H.; Kim, M.N. & Lee, S.H. (2002). Microvibration transducer using silicon elastic body for an implantable middle ear hearing aid, *Mechatronics*, Vol.12, pp.1173-1184.

Park, S.N.; Kim, J.K. & Suh, H. (2004). Evaluation of antibiotic-loaded collagen-hyaluronic acid matrix as a skin substitute, *Biomaterials*, Vol.25, pp.3689-3698.

Park, S.N.; Lee, H.J.; Lee, K.H. & Suh, H. (2003). Biological characterization of EDC-crosslinked collagen-hyaluronic acid matrix in dermal tissue restoration, *Biomaterials*, Vol.24, pp.1631-1641.

Patankar, M,S.; Oehninger, S.; Barnett, T.; Williams, R.L. & Clark, G.F. (1993). A revised structure for fucoidan may explain some of its biological activities, *Journal of Biological Chemistry*, Vol. 268, No.29, pp.21770-21776.

Paul, W. & Sharma, C.P. (2004). Chitosan and alginate wound dressings: a short review, *Trends in Biomaterials & Artificial Organs*, Vol.18, pp.18-23.

Peng, H.T. & Shek, P.N. (2009). Development of in situ-forming hydrogels for hemorrhage control, *Journal of Materials Science: Materials in Medicine*, Vol.20, No.8, pp.1753-1762.

Peters, G. & Wirth, CJ. (2003). The current state of meniscal allograft transplantation and replacement, *Knee*, Vol.10, No.1, pp.19-31.

Pietramaggiori, G.; Yang, H.J.; Scherer, S.S.; Kaipainen, A.; Chan, R.K.; Alperovich, M.; Newalder, J.; Demcheva, M.; Vournakis, J.N.; Valeri, C.R.; Hechtman, H.B. & Orgill, D.P. (2008). Effects of poly-N-acetyl glucosamine (pGlcNAc) patch on wound healing in db/db mouse, *Journal of Trauma*, Vol.64, No.3, pp.803-808.

Price, R.D.; Das-Gupta, V.; Frame, J.D. & Navsaria, H.A. (2001). A study to evaluate primary dressings for the application of cultured keratinocytes, *British Journal of Plastic Surgery*, Vol.54, pp. 687-696.

Pruitt, B.A. & Levine, N.S. (1984). Characteristics and uses of biologic dressings and skin substitutes, Archives of Surgery, Vol.119, pp.312-322.

Rao, S.B. & Sharma, C.P. (2007). Use of chitosan as a biomaterials: studies on its safety and hemostatic potential, *Journal of Biomedical Materials Research*, Vol.34, pp.21-28.

Raphael, K.G.; Marbach, J.J.; Wolford, L.M.; Keller, S.E. & Bartlett, J.A. (1999). Self-reported systemic, immune-mediated disorders in patients with and without proplast-teflon implants of the temporomandibular joint, *Journal of Oral and Maxillofacial Surgery*, Vol.57, pp.364-370.

Ravishanker, R.; Bath, A.S. & Roy, R. (2003). Amnion Bank the use of long term glycerol preserved amniotic membranes in the management of superficial and superficial partial thickness burns, *Burns*, Vol. 29, No.4, pp.69-74.

Ruszczak, Z. (2003). Effect of collagen matrices on dermal wound healing, *Advanced of Drug Delivery Reviews*, Vol.55, pp.1595-1611.

Saliba, M.J. (2001). Heparin in the tratment of burns: a review, *Burns*, Vol.27, pp.349-358.

Sezer, A.D. (2011). Chitosan: properties and its pharmaceutical and biomedical aspects. In: Focus on chitosan research, Samuel P. Davis, editor pp.377-398, Nova Publisher, ISBN: 978-1-61324-454-8, Hauppauge, Newyork, USA.

Sezer, A.D. & Akbuğa, J. (2006). Fucosphere--new microsphere carriers for peptide and protein delivery: preparation and in vitro characterization, *Journal of Microencapsulation*, Vol.23, No.5, pp.513-522.

Sezer, A.D. & Akbuğa, J. (2009). Comparison on in vitro characterization of fucospheres and chitosan microspheres encapsulated plasmid DNA (pGM-CSF): formulation design and release characteristics, *AAPS PharmSciTech*, Vol.10, No.4, pp.1193-1199.

Sezer, A.D. & Cevher, E. (2011). Fucoidan: is a versatile biopolymer for biomedical applications, In: Biomaterials and nanostructures are for active implants, Meital Zilberman, editor. pp.377-406, Springer-Verlag Publisher, ISBN: 978-3-642-18064-4, Heidelberg, Berlin, Germany.

Sezer, A.D.; Cevher, E.; Hatipoğlu, F.; Oğurtan, Z.; Baş, A.L. & Akbuğa, J. (2008). Preparation of fucoidan-chitosan hydrogel and its application as burn healing accelerator on rabbits, *Biological and Pharmaceutical Bulletin*, Vol.31, No.12, pp.2326-2333.

Sezer, A.D.; Cevher, E.; Hatipoğlu, F.; Oğurtan, Z.; Baş, A.L. & Akbuğa, J. (2008). The use of fucosphere in the treatment of dermal burns in rabbits, *European Journal of Pharmaceutics and Biopharmaceutics*, Vol.69, No.1, pp.189-198.

Sezer, A.D.; Hatipoğlu, F.; Cevher, E.; Oğurtan, Z.; Baş, A.L. & Akbuğa, J. (2007). Chitosan films containing fucoidan as a wound dressing for dermal burn healing: Preparation and in vitro/in vivo evaluation, *AAPS PharmSciTech*, Vol.8, No.2, Article 39, E1-E8.

Shakespeare, P. & Shakespeare V. (2002). Survey: use of skin substitute materials in UK burn treatment centres, *Burns*, Vol.28, pp.295-297.

Shakespeare, P. (2001). Burn wound healing and skin substitutes, *Burns*, Vol.27, pp.517-522.

Shen, X.; Nagai, N.; Murata, M.; Nishimura, D.; Sugi, M. & Munekata, M. (2008). Development of salmon milt DNA/salmon collagen composite for wound dressing, *Journal of Materials Science: Materials in Medicine*, Vol.19, pp.3473–3479.

Sheridan, R.L. & Tompkins, R.G. (1999). Skin substitutes in burns, *Burns*, Vol.25, pp.97-103.

Sheridon, R.L.; Morgan, J.R. & Mohammad R. (2001). Biomaterials in burn and wound dressing. Ed: Severian D., Polymeric Biomaterials, Dumitriu Severian, editor. pp.451-458, Marcel Dekker, ISBN: 0-8247-8969-5, New York, USA.

Singer, A.J.; Mohammad, M.; Thode, H.C. & Mcclain, S.A. (2000). Octylcyanoacrylate versus polyurethane for treatment of burns in swine: a randomized trail, *Burns*, Vol.26, pp.388-392.

Skin. Ed: Junqueria, L.C.; Carneiro, J. & Kelley, O.R., Basic Histology. Vol. 18, pp.357-370, Prentice-Hall International Inc., Rio de Janerio, Brasil.

Skin. Ed: Young, B. & Heath J.W., Wheather's Functional Histology. Vol. 9, pp.157-171, Churchill Livingstone, Philadelphia, USA.

Sparkes, B.G.: Immunological responses to termal injury, *Burns*, Vol.23, pp.106-113.

Stashak, T.S.; Farstvedt, E. & Othic, A. (2004). Update on wound dressings: indications and best use, *Clinical Techniques in Equine Practice*, Vol.3, pp.148-163.

Still, J.; Glat, P.; Silverstein, P.; Griswold, J. & Mozingo, D. (2003). The use of a collagen sponge/living cell composite material to treat donor sites in burn patients, *Burns*, Vol.29, No.8, pp.837-841.

Suh, J.K.F. & Matthew, H.W.T. (2000). Application of chitosan-based polysaccharide biomaterials in cartilage tissue engineering: a review, *Biomaterials*, Vol.21, pp.2589-2598.

Sung, J.H.; Hwang, M.R.; Kim, J.O.; Lee, J.H.; Kim, Y.I.; Kim, J.H.; Chang, S.W.; Jin, S.G.; Kim, J.A.; Lyoo, W.S.; Han, S.S.; Ku, S.K.; Yong, C.S. & Choi, H.G. (2010). Gel characterisation and in vivo evaluation of minocycline-loaded wound dressing with enhanced wound healing using polyvinyl alcohol and chitosan, *International Journal of Pharmaceutics*, Vol.392, No.1-2, pp.232-240.

Suzuki, S.; Matsuda, K.; Isshiki, N.; Tamada, Y. & Ikada, Y. (1990). Experimental study of a newly developed bilayer artificial skin, *Biomaterials*, Vol.11, pp.356-360.

Şenel, S. & McClure, S.J. (2004). Potential applications of chitosan in veterinary medicine, *Advanced of Drug Delivery Reviews*, Vol.56, pp.1467-1480.

Şenyuva, C.; Yücel, A.; Erdamar, S.; Çetinkale, O.; Seradjmir, M. & Özdemir, C. (1997). The fate of alloplastic materials placed under a burn scar: an experimental study, *Burns*, Vol.23, pp.484-489.

Tan, W.; Krishnaraj, R. & Desai, T.A. (2001). Evaluation of nanostructured composite collagen-chitosan matrices for tissue engineering, *Tissue Engineering*, Vol.7, pp.203-210.

Tanigawa, J.; Miyoshi, N. & Sakurai, K. (2008). Characterization of chitosan/citrate and chitosan/acetate films and applications for wound healing, *Journal of Applied Polymer Science*, Vol.110, pp.608–615.

The skin and its appendages (The Integument). (1981). Ed: Leeson T.S. & Leeson C.R., Histology. Vol. 10, pp. 308-327, W.B. Saunders Company, Philadelphia, USA.

The skin and its appendages (The Integument). (1988). Ed: Leeson T.S., Leeson C.R., Paparo A.A., Text/Atlas of Histology. Vol. 10, s. 362-393, W.B. Saunders Company, Philadelphia, USA.

Thomas, S. (2000a). Alginate dressings in surgery and wound management-Part 1, *Journal of Wound Care*, Vol. 9 No.2 pp.56-60.

Thomas, S. (2000b). Alginate dressings in surgery and wound management-Part 2, *Journal of Wound Care*, Vol. 9 No.3 pp.115-119.

Thomas, S. (2000c). Alginate dressings in surgery and wound management-Part 3, *Journal of Wound Care*, Vol. 9 No.4 pp.163-166.

Tissue repair and wound healing. (1998). Ed: Porth C.M., Pathophysiology. Vol. 2, pp.43-48, Lippincott, Philadelphia, USA.

Tissue repair and wound healing. (1998). Ed: Porth C.M., Pathophysiology., Vol. 2, pp.289-293, Lippincott, Philadelphia, USA.

Trumble, D.R.; McGregor, W.E. & Magovern, J.A. (2002). Validation of a bone analog model for studies of sternal closure, *Annals of Thoracic Surgery*, Vol.74, pp.739-745.

Ueno, H.; Mori, T. & Fujinaga, T. (2001). Topical formulation and wound healing applications of chitosan. *Advanced of Drug Delivery Reviews*, Vol.52, pp.105-115.

Ueno, H.; Yamada, H.; Tanaka, I.; Kaba, N.; Matsuura, M.; Okumurai, M.; Kadosawa T. & Fujinaga, T. (1999). Accelerating effects of chitosan for healing at, early phase of experimental open wound in dogs, *Biomaterials*, Vol.20, pp.1407-1414.

Uppal, R.; Ramaswamy, G.N.; Arnold, C.; Goodband, R. & Wang, Y. (2011). Hyaluronic acid nanofiber wound dressing: Production, characterization, and in vivo behavior, *Journal of Biomedical Materials Research Part B: Applied Biomaterials*, Vol.97B, pp.20-29.

Van den Kerckhove, E.; Stappaerts, K.; Boeckx, W.; Van den Hof B.; Monstrey, S.; Van der Kelen, A. & Cubber J. (2001). Silicones in the rehabilitation of burns: a review and overview, *Burns*, Vol.27, pp. 205-214.

Van Zuijlen, P.P.M.; Ruurda, J.J.B.; Van Veen, H.A.; Marle, J.V.; Van Trier, A.J.M.; Groenevelt, F.; Kreis, R.W. & Middelkoop, E. (2003). Collagen morphology in human skin and scar tissue: no adaptations in response to mechanical loading at joints, *Burns*, Vol.29, pp.423-431.

Verma, P.R.P. & Iyer, S.S. (2000). Controlled transdermal delivery of propranolol using HPMC matrices: design and *in-vitro* and *in-vivo* evaluation, *Journal of Pharmacy and Pharmacology*, Vol.52, pp.151-156.

Wang, W. (2006). A novel hydrogel crosslinked hyaluronan with glycol chitosan, *Journal of Materials Science: Materials in Medicine*, Vol.17, No.12, pp.1259-1265.

Weinstein-Oppenheimer, C.R.; Aceituno, A.R.; Brown, D.I.; Acevedo, C.; Ceriani, R.; Fuentes, M.A.; Albornoz, F.; Henríquez-Roldán C.F.; Morales, P.; Maclean, C.; Tapia, S.M. & Young, M.E. (2010). The effect of an autologous cellular gel-matrix integrated implant system on wound healing, *Journal of Translational Medicine*, Vol.8, No.59, pp.1-11.

Whelan, J. (2002). Smart bandages diagnose wound infection, *DDT*, Vol.7, pp.9-10.

Whitney, J.D. & Wickline, M.M. (2003). Treating chronic and acute wounds with warming: review of the science and practice implications, *Journal of Wound Care*, Vol.30, pp.199-209.

Winter, G.D. (1965). A note on wound healing under dressings with special reference to perforated-film dressings, *Journal of Investigative Dermatology*, Vol.45, pp.299-302.

Wright, K.A.; Nadire, K.B.; Busto, P.; Tubo, R.; McPherson, J.M. & Wentworth, B.M. (1998). Alternative delivery of keratinocytes using a polyurethane membrane and the implications for its use in the treatment of full-thickness burn injury, *Burns*, Vol.24, pp.7-17.

Xu, C.; Xu, F.; Wang, B. & Lu, T. (2011). Electrospinning of poly(ethylene-co-vinyl alcohol) nanofibres encapsulated with ag nanoparticles for skin wound healing, *Journal of Nanomaterials*, Article ID 201834.

Xu, H.,; Ma, L.; Shi, H.; Gao, C. & Han, C. (2007). Chitosan–hyaluronic acid hybrid film as a novel wound dressing: *in vitro* and *in vivo* studies, *Polymers for Advanced Technologies*, Vol.18, pp.869–875.

Yenerman, M. (1986). Genel Patoloji. 1st edition, pp.271-294, Istanbul Üniversitesi, Istanbul, Turkey.

Yoo, H. & Kim, H. (2008). Characteristics of Waterbornepolyurethane/poly (nvinylpyrrolidone) composite films for wound-healing dressings, *Journal of Applied Polymer Science*, Vol.107, pp.331–338.

7

Extracellular Matrix Adjuvant for Vaccines

Mark A. Suckow[1], Rae Ritchie[2] and Amy Overby[2]
[1]University of Notre Dame;
[2]Bioscience Vaccines, Inc.
United States of America

1. Introduction

In 1796, Edward Jenner first used the term, "vaccination," to describe his studies which used poxvirus derived from lesions in cows to protect humans against infection with smallpox (Barquet & Domingo, 1997). Later, Louis Pasteur demonstrated that animals and people could be protected against disease when administered microbes that had been attenuated to reduce pathogenicity. From this early work, it became evident that stimulation of the immune system by exposure to specific antigens associated with pathogens could lead to a response that would protect individuals from infection and, of paramount importance, disease associated with infection.

Since the times of Jenner and Pasteur, vaccination has proved over time to be one of the most cost-effective means to control infectious disease (Kaufmann, 2007). For example, in humans smallpox has been eradicated, and the incidence of polio has been greatly reduced; and in cattle, rinderpest has been eradicated (Kieny & Girard, 2005; Normile, 2008). Indeed, stimulation of the immune system by vaccination has proved to be a cornerstone of preventive medical strategies for several decades in both human and veterinary medicine.

The targets within a vaccine, and against which the immune response is directed, are referred to as "antigens." Often, antigens consist of proteins, though some vaccines utilize polysaccharides, nucleic acids, toxoids, peptides, and inactivated whole or fractions of microorganisms or cells as antigens (Liljeqvist & Ståhl, 1999). Great effort has been given to identification and production of purified recombinant protein subunit vaccines as a way to drive the immune system to target specific antigens key to the colonization, survival, and pathogenesis of infectious agents. Similar work has recently extended to the use of vaccination as an approach to cancer treatment, with vaccines based upon antigens ranging from recombinant subunit proteins to whole, inactivated cancer cells being evaluated in preclinical models and, in some cases, clinical trials (Buonaguro et al., 2011; Melenhorst & Barrett, 2011). In spite of significant progress in identification and production of vaccine antigens, many antigens stimulate only a weak immune response insufficient to offer protection to the patient.

2. Vaccine adjuvants

Adjuvants are compounds added to vaccines to improve the immune response to vaccine antigens. Indeed, the word 'adjuvant' derives from the Latin 'adjuvans' which means 'to help.' Adjuvants exert their action in a number of different ways, such as by increasing the immunogenicity of weak antigens; strengthening the immune responses of individuals with weak immune systems; reducing the amount of antigen needed in the vaccine, thus reducing the cost; extending the duration of the immune response; modulating antibody avidity, specificity, or isotype distribution; and promoting specific forms of immunity, such as humoral, cell-mediated, or mucosal immunity.

2.1 Types of vaccine aduvants

A variety of compounds have been investigated for potential use as vaccine adjuvants. For example, immunostimulatory molecules such as the cytokines IL-2, IL-12, gamma-interferon, and granulocyte-macrophage colony-stimulating factor (GM-CSF) have all been shown to enhance aspects of the immune response following vaccination (Coffman et al., 2010). However, because these molecules are proteins that are relatively expensive, have short half-lives *in vivo*, and may exert a variety of other, often unpredictable systemic effects, their practical application as vaccine adjuvants is unlikely.

Compounds which lack the challenges that restrict the application of cytokines, but which in turn enhance immunostimulatory cytokines, have also been studied. Monophosphoryl lipid A (MPL), a molecule derived from bacterial lipopolysaccharide (LPS), stimulates the release of cytokines, likely through interaction with toll-like receptor 4. MPL induces the synthesis and release of IL-2 and gamma-interferon (Gustafson & Rhodes, 1992; Ulrich & Myers, 1995). In contrast, saponins are derived from the bark of a Chilean tree, *Quillaja saponaria*. The saponin QS21 is a potent adjuvant for IL-2 and gamma-interferon; and it appears to act by intercalating into cell membranes through interaction with structurally similar cholesterol. This interaction results in formation of pores in the cell membrane, and it is thought that this may allow antigens a direct pathway for presentation to the immune system (Glaueri et al., 1962). QS21 has been associated with pain on injection and local reactions, and the balance between potency and adverse events is an important consideration (Kensil & Kamer, 1998). More recently, unmethylated CpG dinuclelotides have been evaluated for potential use as vaccine adjuvants. Such CpG molecules are common in bacterial DNA, but not in vertebrate DNA, and it is thought that the interaction of the immune system with these moieties is related to an evolutionary characteristic of vertebrate defense against microbial infection (McKluskie & Krieg, 2006). In this way, then, it may be that cells of the immune system are particularly adept at recognizing antigens conjugated to CpG DNA. Responses to CpG DNA are mediated by binding to toll-like receptor 9 (Hemmi et al., 2000). Further, it is believed that CpG conjugates are taken up by non-specific endocytosis and that endosomal maturation is necessary for cell activation and release of pro-inflammatory cytokines (Sparwasser, et al., 1998).

One approach toward vaccine adjuvants has been to focus on delivery of antigens to the immune system. In this regard, a variety of methods for delivery of antigens have been investigated, including oil-in-water liposomes, alginate microparticles, and inclusion of antigens in live viral or bacterial vectors. For example, the MF59 squalene oil-in-water emulsion has shown an adjuvant effect in animal models for an influenza vaccine (Cataldo & Van Nest, 1997) and hepatitis B vaccine (Traquina et al., 1996). Microparticles produced by polymerization of alginate by divalent cations have been used to vaccinate animals against bacterial respiratory pathogens (Suckow et al., 1999). Alginate microparticles enhance antibody responses when administered with incorporated antigens by either the subcutaneous or mucosal (intra-nasal or peroral) routes. Adjuvants based upon delivery of antigens via live viral or bacterial vectors have gained recent interest, particularly for stimulation of immunity at mucosal surfaces. For example, modified intestinal pathogens such as *Salmonella typhimurium*, *S. cholerasuis*, *Listeria monocytogenes*, and *Escherichia coli* have all been used to deliver antigens, most often through vaccination at mucosal surfaces (Becker et al., 2008). Similarly, poxvirus and adenovirus have been used as a means of delivering antigens for vaccination purposes (Liu, 2010). A primary advantage of vaccination using a live vector is that the replication of the vector within the host can lead to a heightened and sustained immune response to the vectored antigens.

2.2 Aluminum salt adjuvants

The use of aluminum salts (alum) as vaccine adjuvants to boost the immune response to vaccine antigens has a long record of success over the past 70 years (Clapp et al., 2011). In this regard, both aluminum phosphate and aluminum hydroxide have been used, in varying ratios and concentrations specific to different vaccines. Typically, the antigen is adsorbed to alum through electrostatic charge, and the degree of antigen adsorption by aluminum-containing adjuvants is generally considered to be an important characteristic of vaccines that is related to immunopotentation by the adjuvant. In this regard, the World Health Organization (WHO) requires that at least 80% of the antigens in alum-precipitated diphtheria and tetanus toxoid vaccines be adsorbed (Clapp et al, 2011); however, others have found that there appears to be no correlation between the percentage of antigen adsorbed and subsequent antibody production following vaccination (Clausi et al., 2008). Further, antigen presenting cells have been demonstrated to take up desorbed antigen from the interstial fluid as well as antigen adsorbed to aluminum-containing adjuvants (Iver et al., 2003); and aluminum-containing adjuvants potentiate the immune response to non-adsorbed protein antigens (Romero-Mendez et al., 2007), suggesting that adsorption may not be a requirement for adjuvancy by alum.

The most widely accepted theory for the mechanism of alum adjuvancy is the repository effect, whereby the antigen adsorbed by the aluminum-containing adjuvant is slowly released after intramuscular or subcutaneous administration (Seeber, et al., 1991). However, it has been demonstrated using ^{28}Al in rabbits that aluminum quickly appears in the interstitial fluid and is rapidly eliminated from the body (Flarend et al., 1997). The best information regarding the mechanism of action for alum suggests that alum activates a complex of proteins, termed the inflammasome, in lymphocytes and that these proteins initiate a complex cascade of actions which result in an enhanced immune response to antigens (Eisenbarth et al., 2008).

At present, aluminum salts are the only adjuvants that are currently approved for use in human vaccines by the United States Food and Drug Administration (FDA). The FDA limits the amount of aluminum in biological products, including vaccines, to 0.85 mg/dose. However, in spite of this regulatory oversight and a widespread acceptance of safety, aluminum adjuvants have been associated with severe local reactions such as erythema, induration, formation of subcutaneous granulomas, and pain. A further major limitation of aluminum adjuvants is their inability to elicit cell-mediated Th1 responses that are required to control most intracellular pathogens such as those that cause tuberculosis, malaria, leishmaniasis, leprosy, and AIDS ; and to elicit Th1 responses typical of immunity to cancer. Additionally, failure of alum to augment the humoral antibody immune response in some circumstances, such as with influenza immunization in the elderly (Parodi et al., 2011), clearly illustrates the need for additional safe and effective vaccine adjuvants.

3. Extracellular matrix

Extracellular matrix (ECM) is a network of macromolecules, largely proteins and polysaccharides, that aggregate to form a complex meshwork that acts as a both physical lattice and a biologically active promoter for the cellular component of tissue. Proteins typically found in ECM include collagen, elastin, fibronectin, and laminin all of which contribute structural strength to the ECM and adhesive attraction for the cells. Complementing these proteins are glycosaminoglycans, a class of polysaccharide which are often found covalently linked to proteins to constitute proteoglycans. The proteoglycans exist as a highly hydrated, gel-like ground substance in which the proteins are embedded. The proteoglycan gel allows diffusion of nutrients and molecular signals important for cell survival and growth. Several glycosaminoglycans are typically present in ECM, including hyaluronan; chondroitin sulfate and dermatan sulfate; heparan sulfate; and keratan sulfate. The glycosaminoglycans are active participants in the maintenance and survival of cells and exhibit a variety of biological functions. For example, hyaluronan is thought to facilitate migration of cells during the process of tissue repair. Further, proteoglycans are believed to have a major role in chemical signaling between cells. An example is heparan sulfate which binds to fibroblast growth factors which subsequently stimulates proliferation by a variety of cells. Heparan sulfate proteoglycans immobilize secreted chemotactic attractants called chemokines on the endothelial surfaces of blood vessels at sites of inflammation. In this way, the chemokines remain there for a prolonged period, stimulating white blood cells to leave the bloodstream and migrate into the inflamed tissue. In short, the extracellular matrix is an active participant in the dynamic process of tissue function.

3.1 Small intestinal submucosa

Small intestinal submucosa (SIS) is a natural, bioactive extracellular matrix that has proven successful as a tissue graft material in a variety of clinical applications related to tissue repair (Ellis, 2007; Mostow et al., 2005). The material is processed, using porcine intestine as a raw material, into an acellular, multilaminar medical-grade material that serves as a bioscaffold for in-growth of, and subsequent incorporation into, normal, repaired tissue (Fig. 1).

SIS has been found to be a suitable material for tissue engineering applications at varied sites, including the lower urinary tract, the body wall, tendon, ligament, flat bone, cutaneous wounds and blood vessels. One common cause of tissue defects that might benefit from augmentation is tumor resection. It is interesting that, in spite of its tissue growth promoting characteristics, SIS has been shown not to promote re-growth of tumors when added to the remaining tumor bed following surgical tumor resection in a pre-clinical model (Hodde et al., 2004).

It is known that bioactive growth factors are retained in the SIS following lyophilization and sterilization. These growth factors include transforming growth factor-beta (TGF-β), which is important in wound healing, and the highly angiogenic growth factor, basic fibroblast growth factor (FGF-2). In addition, it has been demonstrated that SIS contains glycosaminoglycans such as hyaluronic acid, dermatan sulfate, chondroitin sulfate A, and heparan sulfate (Hodde et al., 1996). Because some of the gylcosaminoglycans endogenous to SIS assist chemotactic cytokines, it is not surprising that antigen-processing cells such as macrophages are drawn to sites where SIS has been implanted (Badylak et al., 2008).

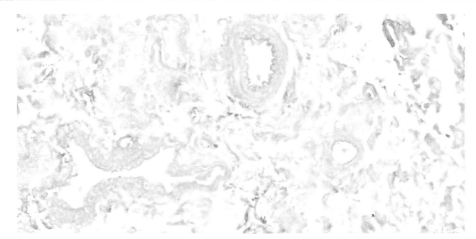

Fig. 1. Photomicrograph of small intestinal submucosa, demonstrating lack of cellular components but with remnant acellular matrix and vascular structures.

4. Use of SIS as a vaccine adjuvant and immune modulator

Because small intestinal submucosa has biological activities that are known to include attraction of antigen-presenting cells, such as macrophages, to sites of implantation, the potential application of the material as a vaccine adjuvant is obvious. That SIS is available as a medical-grade material in sheet form (Fig. 2) and has been safely used in a variety of applications in a large number of patients supports the idea that it could be safely applied as a vaccine adjuvant as well.

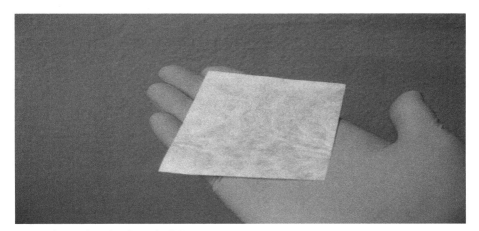

Fig. 2. A sheet of medical-grade SIS.

4.1 SIS enhances the performance of whole cell cancer vaccines
The idea that cancer is a disease that can be treated by vaccination is relatively new, and great effort has been given to development of therapeutic, and in some cases prophylactic,

vaccines. For example, the Provenge® dendritic cell-based vaccine for treatment of prostate cancer was recently approved by the U. S. Food and Drug Administration (Madan & Gulley, 2011); and the Gardasil® and Cervarix® vaccines for prevention of cervical cancer are approved as well (Harper, 2009). Indeed, it is believed by many that the age of cancer immunotherapy has dawned and offers entirely new possibilities for the clinical approach to cancer.

The use of whole tumor cells as vaccine components allows a greatly increased menu of antigens to be presented to the immune system. Although many antigenic moieties in such vaccine preparations may be unidentified, it can be presumed that the rich choice of antigenic targets facilitates the likelihood of a successful immune response. As an example, allogeneic (from the same species) human prostate cancer cells have been examined for potential use in vaccination therapy of prostate cancer patients. The basic reasoning for this approach is that because tumor antigens are often conserved between tumors, allogeneic vaccines might stimulate cross-protective immunity. To test this idea, monthly intradermal administrations of a vaccine composed of three inactivated (irradiated) allogeneic prostate cancer cell lines were given for 1 year to patients having progressive disease as defined by two consecutive increases in prostate-specific antigen (PSA). The treatment did not produce any evidence of toxicity and resulted in decreased PSA velocity, as well as a cytokine response profile consistent with a Th1 immune response (Simons et al., 1999). In addition, median time to disease progression was 58 weeks in vaccinated patients compared with 28 weeks for historical controls.

Tissue vaccines represent an expansion on the idea of whole cell vaccines. Tissue vaccines are produced directly from harvested tumor material and do not undergo any *in vitro* culture (Suckow et al., 2007a). In this regard, tissue vaccines include an enormous menu of antigen targets composed of various stages of an evolving, growing population of cancer cells; antigens associated with an evolving and expanding extracellular matrix; and antigens expressed uniquely *in vivo* but not *in vitro*. In the Lobund-Wistar rat model of hormone-refractory prostate cancer, a tissue vaccine was demonstrated to reduce the incidence of autochthonous prostate cancer by 90% (Suckow et al., 2005); reduce metastasis to the lungs by 70% (Suckow et al., 2008a); and augment tumor reduction by external beam irradiation by approximately 50% (Suckow et al., 2008b). Further, a xenogeneic tissue vaccine was demonstrated to reduce growth by 70% of tumors associated with human prostate cancer cells xenotransplanted in immunodeficient mice (Suckow et al., 2007b).

4.1.1 SIS as an adjuvant for prostate cancer whole cell vaccines

Melanoma is a cancer of pigment-producing skin cells referred to as melanocytes. Often aggressive with metastasis to the lymph nodes, lungs, liver, and brain, melanoma is usually first treated by surgical resection of the primary tumor. Though often curative, return of the tumor is not infrequent and can result in a particularly aggressive form of cancer.

To evaluate the ability of SIS to enhance the protective effect conferred by vaccination with a whole cell melanoma vaccine, C57Bl6/J mice were administered B16 mouse melanoma cells to produce subcutaneous tumors. When the tumors were palpable they were harvested, dissociated with a 80-mesh stainless steel screen and allowed to incubate on 2 x 2 cm sections of sterile SIS for three days. Sections of SIS with cells were then treated with 2.5% glutaraldehyde to produce a tissue vaccine as previously described (Suckow et al., 2008c). A separate portion of cells was used to make a tissue vaccine without added SIS, and separate sections of SIS were treated with glutaraldehyde and washed to produce an SIS control treatment.

Groups of naïve mice were administered B16 cells to produce subcutaneous melanoma tumors. Fourteen days after administration of cells, all mice had palpable tumors and were prepared for aseptic surgical debulking of the tumor masses. Mice were anesthetized with a mixture of ketamine hydrochloride, acepromazine maleate, and xylazine; the hair overlying the tumor mass was shaved; and the skin scrubbed with an iodophore. Following incision of the skin, tumors the tumors were carefully dissected free of attachments except for a residual portion of the underlying tumor bed. Groups of mice then either received no further treatment other than wound closure and routine post-surgical care; direct administration onto the tumor bed of 1×10^6 glutaraldehyde-fixed tumor (GFT) cells; application onto the tumor bed of glutaraldehyde-fixed SIS (ECM) with no added cells; or application onto the tumor bed of glutaraldehyde-fixed SIS on which tumor cells had been grown and then fixed (GFT/ECM). Fourteen days after surgery, mice were euthanized and necropsied to assess the re-growth of tumors. As demonstrated in Fig. 3, the combination of GFT/ECM reduced the mass of re-grown tumors by approximately 70% compared to all other treatment groups.

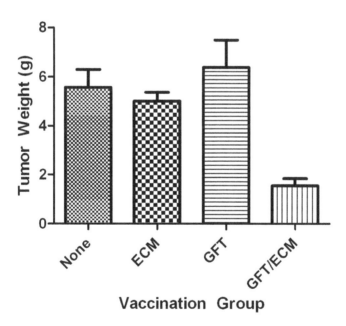

Fig. 3. Neither a melanoma tissue vaccine (GFT) nor SIS (ECM) alone reduced the size of re-grown melanoma tumors compared to controls following surgical resection, however the combination of the tissue vaccine with SIS (GFT/ECM) led to an approximately 70% reduction in tumor mass.

The idea that cancer immunotherapy might be enhanced by inclusion of SIS in the vaccine preparation needed further validation in additional models of cancer. In this regard, prostate cancer was viewed as a likely system for validation, as it is a common cancer and one which has been shown to be responsive to immunotherapy as demonstrated by the Provenge® vaccine. The Lobund-Wistar (LW) rat model of prostate cancer closely replicates the disease in man in that it progresses to a hormone-refractory cancer and readily metastasizes through the circulatory system (Pollard & Suckow, 2005). The LW rat develops autochthonous tumors, and a transplantable cell line (PAIII cells) has been isolated and characterized. When transplanted subcutaneously, PAIII cells rapidly produce aggressive tumors that metastasize to the lungs. Following surgical resection, PAIII tumors become exceptionally aggressive and there is little that can be done to ameliorate consequent growth and spread of the tumor.

As in the mouse melanoma model, subcutaneous tumors were produced as a source of vaccine material. PAIII cells administered subcutaneously rapidly grew into palpable tumors that were harvested 14 days after administration of cells. Following similar procedures as for the melanoma tissue vaccine, harvested tumor cells were added to 2 x 2 cm sections of SIS and allowed to grow for three days at 37° C, followed by glutaraldehyde fixation and extensive washing. Separate groups of rats which were administered PAIII cells 14 days earlier underwent surgical debulking of subcutaneous PAIII tumors then received no treatment other than standard peri-operative care; direct administration onto the tumor bed of 1×10^6 glutaraldehyde-fixed tumor (GFT) cells; application onto the tumor bed of glutaraldehyde-fixed SIS (ECM) with no added cells; or application onto the tumor bed of glutaraldehyde-fixed SIS on which tumor cells had been grown and then fixed (GFT + ECM). Histological examination of a section of the GFT + ECM vaccine material shows that cells readily grow along the edge of SIS (Fig. 4) and within the substance of the SIS (Fig. 5). Rats were euthanized 28 days later and tumors weighed to assess the effect of vaccination on tumor re-growth.

Microscopic examination of SIS upon which harvested tumor tissue was cultured demonstrated robust growth of tumor cells. Cells readily grew along the margin of SIS as shown in Fig. 4 and were arranged as a monolayer along the edge.

Fig. 4. Photomicrograph demonstrating growth of harvested PAIII tumor cells along the edge of SIS.

In contrast, within the substance of the SIS, cells were present as islands of piled-up cells (Fig. 5). Some islands of cells formed what appeared to be primitive vascular structures, consistent with the highly angiogenic character of PAIII tumors.

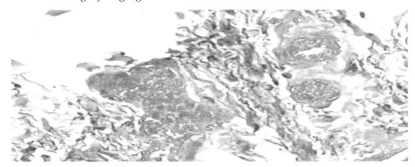

Fig. 5. Photomicrograph demonstrating growth of harvested PAIII tumor cells within the substance of SIS.

The requirement for delivery of nutrients and removal of waste metabolic products characteristic of a rapidly dividing and growing tissue is what drives the need for an expanded blood supply; thus, it is not surprising that cells from a harvested tumor would have an angiogenic phenotype. Indeed, vascular endothelial growth factor (VEGF) has been implicated in a number of aspects of cancer growth, including angiogenesis, remodeling of the ECM, generation of inflammatory cytokines, and hematopoietic stem cell development. Rats vaccinated with the GFT vaccine to which SIS was added had tumors that had a mean weight of 3.91 g, while tumors from rats that had not been vaccinated had a mean weight of 11.63 g (Fig. 6). Similarly, tumors from rats vaccinated with SIS only had a mean weight of 13.04 g and those from rats vaccinated with the GFT vaccine only had a mean weight of 9.96 g. While there was a significant ($P < 0.01$) reduction in tumor weight of rats vaccinated with GFT plus SIS compared to all other groups, tumors from rats of all other treatment groups did not significantly differ in this very aggressive model of cancer (Suckow et al., 2008c).

Adjuvancy of SIS on Post-Resection Vaccines

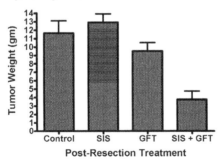

Fig. 6. Neither a PAIII prostate cancer tissue vaccine (GFT) nor SIS (ECM) alone reduced the size of re-grown prostate tumors compared to controls following surgical resection, however the combination of the tissue vaccine with SIS (GFT/ECM) led to an approximately 65% reduction in tumor mass.

Vaccination also reduced the number of rats which had foci of metastasis in the lungs. Specifically, only 40% of rats vaccinated with SIS + GFT had pulmonary metastatic foci, compared to 70% of rats immunized with the GFT vaccine and 100% of rats which were administered only SIS or those receiving no treatment (Fig. 7).

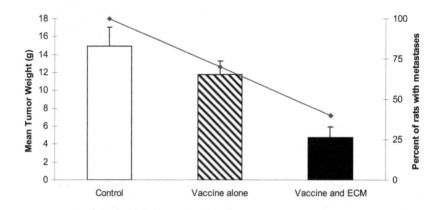

Fig. 7. Reduction in the number of PAIII prostate tumor-bearing rats having pulmonary metastasis (right X-axis). For reference, the mean tumor weight is indicated on the left X-axis.

4.1.2 Gel SIS exhibits adjuvancy for whole cell prostate cancer vaccine

Though vaccines produced from growth of cells directly on SIS extracellular matrix demonstrated remarkable stimulation of anti-tumor responses, it requires surgical implantation. The advantages of a similar vaccine that could be easily and quickly administered are clear. In particular, multiple doses could be given to patients without the requirement for repeated surgery.

Vaccination involving injection into the subcutaneous or intramuscular tissues is rapid and has relatively few adverse outcomes other than acute discomfort at the site of injection. Compounds administered in this way must be of a size and consistency that allow passage through a relatively narrow needle, typically 21-gauge or smaller. Obviously, a sheet of SIS is not amenable to administration in this manner. In this regard, a gel form of SIS, produced by chemical digestion, was developed and evaluated for vaccine adjuvancy. Of note, this same gel preparation has been found to be safe and effective for myocardial infarct repair in rats (Okada et al., 2010).

A tissue vaccine produced from harvested LW rat PAIII prostate tumors, as described earlier, was mixed with SIS gel such that each dose contained 1×10^6 GFT cells. In this case, groups of naïve LW rats were vaccinated with either saline; GFT cells; SIS gel; or GFT cell plus SIS gel. Vaccines were given subcutaneously once before challenge with live PAIII cells, and weekly for two doses afterward. The ability of vaccination to prevent growth of PAIII tumors was assessed by weighing tumors 28 days following PAIII cell challenge. As shown in Fig. 8, neither saline, GFT, nor SIS gel alone demonstrated a protective effect against development of PAIII prostate tumors; however, rats vaccinated with the GFT vaccine combined with SIS gel had tumors which were approximately 75% smaller than those in rats from other vaccination groups.

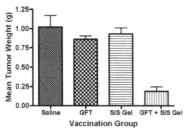

Fig. 8. SIS gel has an adjuvant effect on a tissue vaccine used to prevent prostate cancer.

An obvious concern with use of SIS as an adjuvant for cancer vaccines is whether a material that is known to promote tissue growth, and is used clinically for such applications, would also promote the growth of cancer tissue. Interestingly, when PAIII cells were administered directly onto implanted SIS sections in LW rats, no difference in tumor growth or metastasis was noted compared to control rats that did not have SIS implants (Hodde et al., 2004). This result demonstrates that the SIS extracellular matrix material is capable of promoting growth of normal tissue, but not neoplastic tissue. It may be that the mild inflammatory process which accompanies implantation of SIS facilitates growth of normal tissue, but recognizes foreign material or abnormal cells which happen to be in the vicinity of SIS, thereby alerting the immune system and stimulating an immune response to combat the challenge presented by tumor cells or other pathogenic agents.

4.2 SIS enhances the performance of vaccines for diseases associated with infectious agents

While the idea of vaccination to prevent or treat cancer is relatively new, the concept of vaccination for prevention of diseases associate with infectious agents is not. In 1796, Edward Jenner inoculated an eight-year-old boy with material taken from an active cowpox lesion on another individual and discovered that this strategy could be used to confer protection to smallpox (Barquet & Domingo, 1997). Since that time, vaccines have been developed for a variety of diseases related to infectious agents, including polio, tetanus, influenza, whooping cough (pertussis), measles, diphtheria, and hepatitis (Stern & Markel, 2005).

In spite of the many successes with vaccination, there are some infectious agents for which successful vaccination has proved elusive. For example, malaria and HIV are both important diseases caused by infectious agents and for which there are a lack of viable vaccination strategies. Because SIS proved to be a powerful adjuvant for vaccines designed to protect against cancer, it was reasoned that the material might also have value as an adjuvant for vaccines designed to protect against disease associated with infectious agents.

4.2.1 Particulate SIS vaccine adjuvant

Although sheet SIS and gel SIS were effective as adjuvants for cancer vaccines, it was reasoned that an adjuvant composed of particulate SIS would allow greater surface area of SIS to be exposed and interact with the cellular immune system. This was seen as an advantage because many antigens associated with vaccines for infectious diseases are composed of proteins or sub-units that are more rapidly degraded and processed than whole, inactivated cells (as with whole cell cancer vaccines). Thus, a more rapid and robust adjuvant stimulation was expected to offer greater efficacy for infectious disease vaccines. Particulate SIS was produced by mechanical grinding of medical grade sheet SIS. Sterility of the preparation was maintained and final particulate size was limited to 150 μm. The resulting product was a fine, white dust-like particulate (Fig. 9).

Fig. 9. Particulate SIS. Individual particles in this preparation are no greater than 150 μm in diameter.

4.2.2 Particulate SIS is an effective adjuvant for tetanus vaccine

Though relatively rare in the developed world, tetanus remains an important cause of death worldwide, with up to 1 million deaths estimated annually (Dietz et al., 1996). The disease, characterized by cardiovascular complications which result from autonomic dysfunction, is caused by a Gram-positive bacillus, *Clostridium tetani*. This microorganism is ubiquitous in soil; and under the anaerobic conditions found in necrotic tissue, the bacterium secretes two toxins: tetanolysin, which damages local tissues and optimizes conditions for further bacterial multiplication; and tetanospasmin, which is responsible for development of the clinical disease. Tetanospasmin, also commonly referred to as tetanus toxin, is a two-chain polypeptide of 150,000 Da which, when cleaved by tissue proteases, yields a light chain that acts pre-synaptically to prevent neurotransmitter release (Wright et al., 1989).

Vaccination for the prevention of tetanus has been available since 1923, with routine vaccination beginning later, mostly during the 1950s and 1960s. The vaccine is most commonly a suspension of alum-precipitated (aluminum potassium sulfate) tetanus toxoid. Though vaccination stimulates strong protective immunity in most cases, serological surveys have demonstrated an increasing proportion of patients with inadequate immunity with advancing age (Cook et al., 2001). Though some of these individuals were never vaccinated, a substantial number simply lost immunity over time (Prospero, et al., 1998).

Because immunization with alum-adjuvanted tetanus toxoid is very effective at stimulating protective antibody titers, it represents a good model against which to compare other vaccine adjuvants. In this regard, we undertook studies to evaluate the ability of ECM (SIS) adjuvant to stimulate anti-tetanus toxin antibody and protective responses in a mouse model.

Groups of 15 adult female balb/c mice were vaccinated subcutaneously with 0.03 μg/dose of tetanus toxoid (TT) in sterile saline either with no adjuvant; with 300 μg of alum

(Alhydrogel®); or with 300 µg of particulate ECM. In addition, an untreated control group was included in these studies. Mice were vaccinated initially and received booster vaccinations five weeks later. In addition, serum samples for evaluation of antibody responses were obtained immediately prior to challenge with tetanus toxin, 10 weeks after initial vaccination.

Serum IgG anti-tetanus toxin antibody responses were evaluated by enzyme-linked immunosorbent assay (ELISA). While control animals demonstrated no appreciable anti-tetanus toxin antibody titers, samples from mice vaccinated with alum-adjuvanted tetanus toxoid had measureable antibody levels. In contrast, serum samples from mice vaccinated with tetanus toxoid that was adjuvanted with ECM had five-fold higher titers compared to samples from mice vaccinated with alum-adjuvanted tetanus toxoid (Fig. 10)

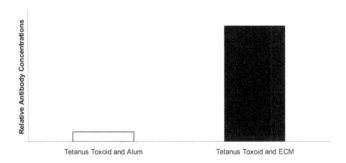

Fig. 10. Relative antibody concentrations in mice immunized with the tetanus toxoid combined with alum or ECM adjuvant.

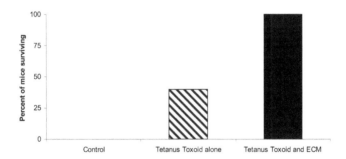

Fig. 11. Survival of mice following challenge with 1 ng/mouse of tetanus toxin

The strength of the anti-tetanus toxin serum antibody response is a reasonable measure of the likely protective effect; however, to assure that vaccination correlated with protection, vaccinated mice were challenged with 1 ng of tetanus toxin administered intraperitoneally. Mice were then observed over the next 96 hours and the number of surviving mice recorded for each group. All mice vaccinated with tetanus toxoid in alhydrogel of ECM survived challenge, while only one-third of mice vaccinated with tetanus toxoid alone, and no untreated control mice survived. Figure 11 shows the percent of surviving mice in mice vaccinated with tetanus toxoid and ECM compared to the non-vaccinated and non-adjuvanted tetanus toxoid-vaccinated control mice.

5. Conclusion

Vaccination has greatly reduced the incidence of many infectious diseases. Currently, a variety of new vaccines based on nucleic acids or other subunits are in development. Further, great promise is offered by vaccination for the treatment and prevention of cancer. With these new vaccines comes a need for improved adjuvants which are safe and effective. Though aluminum salts have been used for adjuvant purposes with great success, there are some vaccines and some patient populations for which alum is not effective.

We have demonstrated the powerful adjuvant effect of SIS, a medical grade ECM. SIS demonstrated a powerful adjuvant effect for vaccines against two types of cancer, melanoma and prostate cancer, and we believe that it will likely prove efficacious with vaccines for other cancers. We further used SIS adjuvant with tetanus toxoid vaccine as a model for evaluation in a vaccine for disease associated with infectious pathogens. We found that the ECM-adjuvanted vaccine stimulated markedly higher anti-tetanus toxin antibody levels than, and offered clinical protection that was at least as good as that conferred by, the alum-adjuvanted vaccine.

In summary, the ECM material, SIS, is an effective vaccine adjuvant and offers an outstanding alternative to the current standard, alum. SIS has a proven safety record when used in humans for a variety of other applications, and we believe that it has unlimited potential for use as a vaccine adjuvant.

6. References

Badylak, S. F., Valentin, J. E., Ravindra, A. K., McCage, G. P., & Stewart-Akers, A. M. (2008). Macrophage phenotype as a determinant of biologic scaffold remodeling. *Tissue Engineering Part A*, Vol. 14, No. 11, (November, 2008), pp. 1835-1842, ISSN 1937-3341.

Barquet, N., & Domingo, P. (1997). Smallpox: the triumph over the most terrible of the ministers of death. *Annals of Internal Medicine*, Vol. 127, No. 8, (October, 1997), pp.635- 642, ISSN 0003-4819.

Becker, P. B., Noerder, M., & Guzman, C. A. (2008). Genetic immunization: bacteria as DNA vaccine delivery vehicles. *Human Vaccines*, Vol. 4, No. 3, (May, 2008), pp. 189-202, ISSN 1554-8600.

Buonaguro, L., Petrizzo, A., Tornesello, M. L., & Buonaguro, F. (2011). Translating tumor antigens into cancer vaccines. *Clinical and Vaccine Immunology*, Vol. 18, No. 1, (January, 2011), pp. 23-34, ISSN 1556-6811.

Cataldo, D. M., & Van Nest, G. (1997). The adjuvant MF59 increases the immunogenicity and protective efficacy of subunit influenza vaccine in mice. *Vaccine*, Vol. 15, No. 16, (November, 1997), pp. 1710-1715, ISSN 0264-410X.

Clapp, T., Siebert, P., Chen, D., & Braun, L. J. (2011). Vaccines with aluminum-containing adjuvants: optimizing vaccine efficacy and thermal stability. *Journal of Pharmaceutical Sciences*, Vol. 100, No. 2, (February, 2011), pp. 388-401, ISSN 1520-1617.

Clausi, A., Cummiskey, J., Merkley, S., Carpenter, J. F., Braun, L. J., & Randolph, T. W. (2008). Influence of particle size and antigen binding on effectiveness of aluminum salt adjuvants in a model lysozyme vaccine. *Journal of Pharmaceutical Sciences*, Vol. 97, No. 12, (December, 2008), pp. 5252-5262, ISSN 1520-1617.

Coffman, R. L., Sher, A., & Seder, R. A. (2010). Vaccine adjuvants: putting innate immunity to work. *Immunity*, Vol. 33, No. 4, (October, 2010), pp. 492-503, ISSN 1074-7613.

Cook, T. M., Protheroe, R. T., & Handel, J. M. (2001). Tetanus: a review of the literature. *British Journal of Anaesthesia*, Vol. 87, No. 3, (September, 2001), pp. 477-487, ISSN 0007- 0912.

Dietz, V., Milestien, J. B., Van Loon, F., Cochi, S., & Bennett, J. (1996). Performance and potency of tetanus toxoid: implications for eliminating neonatal tetanus. *Bulletin of the World Health Organization*, Vol. 74, No. 6, (June, 1996), pp. 619-628, ISSN 0042-9686.

Eisenbarth, S. C., Colegio, O. R., O'Connor, W., Sutterwala, F. S., & Flavell, R. A. (2008). Crucial role for the Nalp3 inflammasome in the immunostimulatory properties of aluminum adjuvants. *Nature*, Vol. 453, No. 7198, (June, 2008), pp. 1122-1126, ISSN 0028-0836.

Ellis, C. N. (2007). Bioprosthetic plugs for complex anal fistulas: an early experience. *Journal of Surgical Education*, Vol. 64, No. 1, (January, 2007), pp. 36-40, ISSN 1931-7204.

Flarend, R. E., Hem, S. L., White, J. L., Elmore, D., Suckow, M. A., Rudy, A. C., & Dandashli, E. (1997). In vivo absorption of aluminum-containing vaccine adjuvants using [26]Al. *Vaccine*, Vol. 15, No. 12-13, (August, 1997), pp. 1314-1318, ISSN 0264-410X.

Glaueri, A. M., Dingle, J. T., & Lucy, J. A. (1962). Action of saponins on biologic membranes. *Nature*, Vol. 196, (December, 1962), pp. 953-959, ISSN 0028-0836.

Gustafson, G. L., & Rhodes, M. J. (1992). Bacterial cell wall products as adjuvants: early interferon gamma as a marker for adjuvants that enhance protective immunity. *Research in Immunology*, Vol. 143, No. 5, (June, 1992), pp. 483-488, ISSN 0923-2494.

Harper, D. M. (2009). Currently approved prophylactic HPV vaccines. *Expert Review of Vaccines*, Vol. 8, No. 12, (December, 2009), pp. 1663-1679, ISSN 1476-0584.

Hemmi, H., Takeuchi, A., Kawai, T., Kaisho, T., Sato, S., Sanjo, H., Matsumoto, M., Hoshino, K., Wagner, H., Takeda, K., Akira, S., & Moingeon, P. (2000). A toll-like receptor recognizes bacterial DNA. *Nature*, Vol. 408, No. 6813, (December, 2000), pp. 740-745, ISSN 0028-0836.

Hodde, J. P., Badylak, S. F., Brightman, A. O., & Voytik-Harbin, S. L. (1996). Glycosaminoglycan content of small intestinal submucosa: a bioscaffold for tissue replacement. *Tissue Engineering*, Vol. 2, No. 3, (Fall, 1996), pp. 209-217, ISSN 2152-4947.

Hodde, J. P., Suckow, M. A., Wolter, W. R., & Hiles, M. C. (2004). Small intestinal submucosa does not promote PAIII tumor growht in Lobund-Wistar rats. *Journal of Surgical Research*, Vol. 120, No. 2, (August, 2004), pp. 189-194, ISSN 0022-4804.

Iver, S., HogenEsch, H., & Hem, S. L. (2003). Relationship between the degree of antigen adsorption to aluminum hydroxide adjuvant in interstitial fluid and antibody production. *Vaccine*, Vol. 21, No. 11-12, (March, 2003), pp. 1219-1223, ISSN 0264-410X.

Kaufmann, S. H. E. (2007). The contribution of immunology to the rational design of novel antibacterial vaccines. *Nature Reviews Microbiology*, Vol. 5, No. 7, (July, 2007), pp. 491- 504, ISSN 1740-1526.

Kensil, C. R., & Kamer, R. (1998). QS-21: a water-soluble triterpine glycoside adjuvant. *Expert Opinion on Investigational Drugs*, Vol. 7, No. 9, (September, 1998), pp. 1475-1482, ISSN 1744-7658.

Kieny, M. P., & Girard, M. P. (2005). Human vaccine research and development: an overview. *Vaccine*, Vol. 23, No. 50, (December, 2005), pp. 5705-5707, ISSN 0264-410X.

Liljeqvist, S., & Stahl, S. (1999). Production of recombinant subunit vaccines: protein immunogens, live delivery systems and nucleic acid vaccines. *Journal of Biotechnology*, Vol. 73, No. 1, (July, 1999), pp. 1-33, ISSN 0717-3458.

Liu, M. A. (2010). Immunologic basis of vaccine vectors. *Immunity*, Vol. 33, No. 4, (October, 2010), pp. 504-515, ISSN 1074-7613.

Madan, R. A., & Gulley, J. L. (2011). Sipleucel-T: harbinger of a new age of therapeutics for prostate cancer. *Expert Review of Vaccines*, Vol. 10, No. 2, (February, 2011), pp. 141-150, ISSN 1476-0584.

McKluskie, M. L., & Krieg, A. M. (2006). Enhancement of infectious disease vaccines through TLR9-dependent recognition of CpG DNA. *Current Topics in Microbiology and Immunology*, Vol. 311, pp. 155-178, ISSN 0070-217X.

Melenhorst, J. J., & Barrett, A. L. (2011). Tumor vaccines and beyond. *Cytotherapy*, Vol. 13, No. 1, (January, 2011), pp. 8-18, ISSN 1465-3249.

Mostow, E. N., Haraway, G. D., Dalgin, M. Hodde, J. P., & King, D. (2005). Effectiveness of an extracellular matrix graft (OASIS Wound Matrix) in the treatment of chronic leg ulcers: a randomized clinical trial. *Journal of Vascular Surgery*, Vol. 41, No. 5, (May, 2005), pp. 837-843, ISSN 0741-5214.

Normile, D. (2008). Rinderpest driven to extinction. *Science*, Vol. 319, No. 5870, (March, 2008), pp. 1606-1609, ISSN 1095-9203.

Okada, M., Payne, T. R., Oshima, H., Momoi, N., Tobita, K., & Huard, J. (2010). Differential efficacy of gels derived from small intestinal submucosa as an injectable biomaterial for myocardial infarct repair. *Biomaterials*, Vol. 31, No. 30, (October, 2010), pp. 7678- 7683, ISSN 0142-9612.

Parodi, V., de Florentis, D., Martini, M., & Ansaldi, F. (2011). Inactivated influenza vaccines: recent progress and implications for the elderly. *Drugs & Aging*, Vol. 28, No. 2, (February, 2011), pp. 93-106, ISSN 1170-229X.

Pollard, M., & Suckow, M. A. (2005). Hormone-refractory prostate cancer in the Lobund-Wistar rat. *Experimental Biology & Medicine*, Vol. 230, No. 8, (September, 2005), pp. 520- 526, ISSN 1535-3702.

Prospero, E., Appignanesi, R., E'Errico, M. M., & Carle, F. (1998). Epidemiology of tetanus in the Marshes Region of Italy. *Bulletion of World Health Organization*, Vol. 76, No. 1, (January, 1998), pp. 47-54, ISSN 0042-9686.

Romero-Mendez, I. Z., Shi, Y., HogenEsch, H., and Hem, S. L. (2007). Poentiation of the immune response to non-adsorbed antigens by aluminum-containing adjuvants. *Vaccine*, Vol. 25, No. 5 (January, 2007), pp. 825-833, ISSN 0264-410X.

Seeber S. J., White, J. L., & Hem, S. L. (1991). Solubilization of aluminum-containing adjuvants by constituents of interstitial fluid. *Journal of Parenteral Science and Technology*, Vol. 45, No. 3, (May, 1991), pp. 156-159, ISSN 0279-7976.

Simons, J. W., Mikhak, B., Chang, J. F., DeMarzo, A. M., Carducci, M. A., Lim, M., Weber, C. E., Baccala, A. A., Goemann, M. A., Clift, S. M., Ando, D. G., Levitsky, H. I., Cohen, L. K., Sanda, M. G., Mulligan, R. C., Partin, A. W., Carter, H. B., Piantadosi, S., Marshall, F. F., & Nelson, W.G. (1999). Induction of immunity to prostate cancer antigens: results of a clinical trial of vaccination with irradiated autologous prostate tumor cells engineered to secrete granulocyte-macrophage colony-stimulating factor using ex vivo gene transfer. *Cancer Research*, Vol. 59, No. 20, (October, 1999), pp. 5160- 5168, ISSN 1010-4283.

Sparwasser, T., Koch, E. S., Vabulas, R. M., Heeg, K., Lipford, G. B., Ellwart, J. W., & Wagner, H. (1998). Bacterial DNA and immunostimulatory CpG oligonucleotides trigger maturation and activation of murine dendritic cells. *European Journal of Immunology*, Vol. 28, No. 6, (June, 1998), pp. 2045-2054, ISSN 0014-2980.

Stern, A. M., & Markel, H. (2005). The history of vaccines and immunization: familiar patterns, new challenges. *Health Affairs*, Vol. 24, No. 3 (May, 2005), pp. 611-621, ISSN 0278-2715.

Suckow, M. A., Hall, P., Wolter, W., Sailes, V., & Hiles, M. C. (2008c). Use of an extracellular matrix material as a vaccine carrier and adjuvant. *Anticancer Research*, Vo. 28, No. 5, (September, 2008), pp. 2529-2534, ISSN 0250-7005.

Suckow, M. A., Heinrich, J., & Rosen, E. D. (2007). Tissue vaccines for cancer. *Expert Review of Vaccines*, Vol. 6, No. 6, (December, 2007), pp. 925-937, ISSN 1476-0584.

Suckow, M. A., Rosen, E. D., Wolter, W. R., Sailes, V., Jeffrey, R., & Tenniswood, M. (2007). Prevention of human PC-346C prostate cancer growth in mice by a xenogeneic vaccine. *Cancer Immunology & Immunotherapy*, Vol. 56, No. 8, (January, 2007), pp. 1275-1283, ISSN 0340-7004.

Suckow, M. A., Siger, L., Bowersock, T. L., Turek, J. J., Van Horne, D., Borie, D., Taylor, A., Park, H., & Park, K. (1999). *Polysaccharide Applications: Cosmetics and Pharmaceuticals* (1st Edition), American Chemical Society, ISBN 0-8412-3641-0, Washington, D. C.

Suckow, M. A., Wheeler, J. D., Wolter, W. R., Sailes, V., & Yan, M. (2008b) Immunization with a tissue vaccine enhances the effect of irradiation of prostate tumors. *In Vivo*, Vol. 22, No. 2, (March, 2008), pp. 171-177, ISSN 0258-851X.

Suckow, M. A., Wolter, W. R., & Pollard, M. (2005). Prevention of autochthonous prostate cancer by immunization with tumor-derived vaccines. *Cancer Immunology & Immunotherapy*, Vol. 54, No. 6, (June, 2005), pp. 571-576, ISSN 0340-7004.

Suckow, M. A., Wolter, W. R., & Sailes, V. T. (2008a). Inhibition of prostate cancer metastasis by administration of a tissue vaccine. *Clinical & Experimental Metastasis*, Vol. 25, No. 8, (August, 2008), pp. 913-918, ISSN 0262-0898.

Traquina, P., Morandi, M., Contorni, M., & Van Nest, G. (1996). MF59 adjuvant enhances the antibody response to recombinant hepatitis B surface antigen vaccine in primates. *Journal of Infectious Diseases*, Vol. 174, No. 6, (December, 1996), pp. 1168-1175, ISSN 0022-1899.

Ulrich, J. T., & Myers, K. R. (1995). Monophosphoryl lipid A as an adjuvant: past experiences and new directions. *Pharmaceutical Biotechnology*, Vol. 6, pp. 495-524, ISSN 1078-0467.

Wright, D. K., Lalloo, U. G., Nayaiger, S., & Govender, P. (1989). Autonomic nervous system dysfunction in severe tetanus: current perspectives. *Critical Care Medicine*, Vol. 17, No. 4, (April, 1989), pp. 371-375, ISSN 1530-0293.

8

Cellular Systems and Biomaterials for Nerve Regeneration in Neurotmesis Injuries

Ana Colette Maurício[1,2] et al.*,
[1]Centro de Estudos de Ciência Animal (CECA), Instituto de Ciências e Tecnologias
Agrárias e Agro-Alimentares (ICETA), Universidade do Porto (UP),
[2]Instituto de Ciências Biomédicas Abel Salazar (ICBAS), UP,
Portugal

1. Introduction

A relevant number of peripheral nerve injuries can only be dealt through reconstructive surgical procedures. Despite continuous refinement of microsurgery techniques, peripheral nerve repair still stands as one of the most challenging tasks in neurosurgery. Particularly problematic is the fact that despite the good regenerative ability of peripheral nerves and successful surgical nerve repair functional recovery is most often disappointing in these patients (Lundborg, 2002). While direct nerve repair should be the procedure of choice whenever tension-free suturing is possible, in many cases there is a significant loss of nerve tissue and resulting nerve gap. In these cases a nerve graft might be necessary for adequate nerve repair (Lundborg, 2002). Nerve grafting, however have some disadvantages, the most prominent being donor site morbidity that may lead to a secondary sensory deficit and occasionally neuroma and pain. In addition, non-matching donor and recipient nerve diameters often occur, which might be at the basis of poor functional recovery (May, 1983). Entubulation offers advantages over autographs, including the potential to manipulate the regeneration environment within the tube-guide (Fields et al., 1989). Consequently, guidance of regenerating axons is not only achieved by a mechanical effect but also by a chemical effect (such as accumulation of neurotrophic factors) (Meek & Coert, 2002). Nerve guides can be made of biological or synthetic materials and, among the latter, both non-absorbable (e.g. silicon) and biodegradable tubes have been used (Schmidt & Leach, 2003). Biodegradable nerve guides must be preferred since no foreign body material will be left in

*Andrea Gärtner[1,2], Paulo Armada-da-Silva[3], Sandra Amado[3], Tiago Pereira[1,2], António Prieto Veloso[3], Artur Varejão[4], Ana Lúcia Luís[1,2] and Stefano Geuna[5]
[1]Centro de Estudos de Ciência Animal (CECA), Instituto de Ciências e Tecnologias Agrárias e Agro-Alimentares (ICETA), Universidade do Porto (UP), Portugal.
[2]Instituto de Ciências Biomédicas Abel Salazar (ICBAS), UP, Portugal.
[3]Faculdade de Motricidade Humana (FMH), Universidade Técnica de Lisboa (UTL), Portugal.
[4]Departamento de Ciências Veterinárias, CIDESD, Universidade de Trás-os-Montes e Alto Douro (UTAD), Portugal
[5]Neuroscience Institute of the Cavalieri Ottolenghi Foundation & Department of Clinical and Biological Sciences, University of Turin, Italy

the host after the device has fulfilled its task. Four types of nerve guides attracted particular attention of our research group: those made of PLGA (Luis et al., 2007b; Luis et al., 2008) and poly-ε-caprolactone (Luis et al., 2008; Varejao et al., 2003a), collagen (Amado et al., 2010) and chitosan (Amado et al., 2008). In particular, chitosan has recently attracted particular attention because of its biocompatibility, biodegradability, low toxicity, low cost, enhancement of wound-healing and antibacterial effects and its potential usefulness in nerve regeneration have been demonstrated both *in vitro* and *in vivo* (Amado et al., 2010; Shirosaki et al., 2005; Simoes et al., 2010). Chitosan is a partially deacetylated polymer of acetyl glucosamine obtained after the alkaline deacetylation of chitin (Senel & McClure, 2004). While chitosan matrices have low mechanical strength under physiological conditions and are unable to maintain a predefined shape after transplantation, their mechanical properties can be improved by modification with a silane agent, namely γ-glycidoxypropyltrimethoxysilane (GPTMS), one of the silane-coupling agents which has epoxy and methoxysilane groups. The epoxy group reacts with the amino groups of chitosan molecules, while the methoxysilane groups are hydrolyzed and form silanol groups, and the silanol groups are subjected to the construction of a siloxane network due to the condensation. Thus, the mechanical strength of chitosan can be improved by the cross-linking between chitosan, GPTMS and siloxane network. We have previously obtained by adding GPTMS, chitosan type III membranes (hybrid chitosan membranes) which have about 110 μm pores and about 90% of porosity, due to the employment of freeze-drying technique, and which were successful in improving sciatic nerve regeneration after axonotmesis (Amado et al., 2008). Wettability of the material surfaces is one of the key factors for protein adsorption, cell attachment and migration (Amado et al., 2008). The addition of GPTMS improves the wettability of chitosan surfaces, and therefore chitosan type III membranes are more hydrophilic than types of chitosan membranes (Shirosaki et al., 2005). In addition, chitosan type III was developed to be more porous, with a larger surface to volume ratio, and to preserve mechanical strength and ability to adapt to different shapes. Significant differences in water uptake between commonly used chitosan and our hybrid chitosan type III were previously reported and it has been shown that they retain about two times as much biological fluid (Tateishi et al., 2002). The Neurolac® and the PLGA90:10 tube-guides guides show several different physical-chemical properties, namely Neurolac® is stiffer than PLGA due to structural reinforcement of the ester bonds and does not degrade so quickly. The biodegradation rate of PLGA tubes is estimated to be around 3 months while that complete resorption of Neurolac® is estimated to be of 16 months approximately (Geuna et al., 2009; Luis et al., 2007b; Luis et al., 2008). The degradation products of Neurolac® are less acidic, which may cause less damage to the surrounding tissue, and they are transparent so that the correct positioning of the nerve stumps may be confirmed after suturing. On the other hand, the biodegradable non woven structure of PLGA tubes may offer some advantages in terms of microporosity which may enhance nerve repair (Luis et al., 2007b; Luis et al., 2008). *In vitro* studies have shown that the four types of tube-guides, namely, the PLGA, the collagen, the hybrid chitosan and the Neurolac® are biocompatible to nerve cells and can facilitate nerve cell attachment, differentiation and growth (Amado et al., 2008; Luis et al., 2007b; Luis et al., 2008; Luis et al., 2008; Wang et al., 2008). Our research group tested the PLGA90:10, the hybrid chitosan type III and the collagen tube-guides covered with the neural cells *in vitro* differentiated for 48 hours in the presence of DMSO (Amado et al., 2010; Amado et al., 2008; Luis et al., 2008; Simoes et al., 2010). The Neurolac® tubes were not tested with the cellular system, since the cell adherence was more difficult to obtain and imply the previous covering with poli-l-lysine (Amado et al., 2010; Amado et al., 2008; Luis et al., 2008; Simoes et al., 2010). It is expected that tissue engineering associating these biomaterials to cellular systems, capable of differentiating into neuron-like cells, may improve peripheral nerve regeneration, in terms of motor, sensory and histomorphometric parameters. The N1E-115 cells derived from mouse neuroblastomas (Amano et al., 1972) can *in vitro* differentiate into neural-cells (Amano et al., 1972; Rodrigues et al., 2005) and were used as a cellular model for stem cells. In addition, mesenchymal cells isolated from the Wharton's jelly of the umbilical cord, were used by our research group (data not published yet). Stem cells can be loosely classified into 3 broad categories based on their time of isolation during ontogenesis: embryonic, fetal and adult (Marcus & Woodbury, 2008). Recent years have witnessed an explosion in the number of adult stem cells populations isolated and characterized. While still multipotent, adult stem cells have long been considered restricted, giving rise only to progeny of their resident tissues. Recently, and currently controversial studies have challenged this dogma, suggesting that adult stem cells may be far more plastic than previously appreciated (Marcus & Woodbury, 2008). Extra-embryonic tissues as stem cell reservoirs offer many advantages over both embryonic and adult stem cell sources. Extra-embryonic tissues, collectively known as the afterbirth, are routinely discarded at parturition, so little ethical controversy attends the harvest of the resident stem cell populations. Most significantly, the comparatively large volume of extra-embryonic tissues and ease of physical manipulation theoretically increases the number of stem cells that can be isolated (Marcus & Woodbury, 2008). The umbilical cord contains two arteries and one vein protected by a proteoglycan rich connective tissue called Wharton's jelly. Within the abundant extracellular matrix of Wharton's jelly resides a recently described stem cell population called Umbilical Cord Matrix Stem Cells. It can be isolated around 400,000 cells per umbilical cord, which is significantly greater than the number of MSCs that can be routinely isolated from adult bone marrow. *In vitro* the Wharton's jelly MSCs cells are capable of differentiation to multiple mesoderm cell types including skeletal muscle and neurons (Fu et al., 2006; Wang et al., 2004). Human MSCs (HMSCs) isolated from Wharton's jelly of the umbilical cord can be easily and ethically obtained and processed compared with embryonic or bone marrow stem cells. These cells may be a valuable source in the repair of the peripheral nerve system. HMSCs from Wharton's jelly of the umbilical cord possess stem cell properties (Yang et al., 2008) and it was previously demonstrated that HMSCs could be induced to differentiate into neuron-like cells (Fu et al., 2006). The transplanted cells were able to promote local blood vessel formation and expression of the neurotrophic factors, brain-derived neurotrophic factor (BDNF) and glial cell line-derived neurotrophic factor (GDNF) (Wang et al., 2004). Therefore, the purpose to cover the PLGA 90:10, hybrid chitosan, and collagen tube-guides used in reconstructed nerves after neurotmesis with a cellular system differentiated *in vitro* into neuronal cells, was to produce locally neurotrophic factors in a physiological concentration. The role of neurotrophic factors in neural regeneration has been the focus of extensive research (Amado et al., 2010; Hu et al., 1997). The influence of these factors in neural development, survival, outgrowth, and branching has been explored on various levels, from the molecular level to the macroscopic tissue responses. Neurotrophic factors promote a variety of neural responses, like the survival and outgrowth of the motor and sensory nerve fibers, and are implicated in spinal cord and peripheral nerve regeneration (Geuna et al., 2009; Johnson et al., 2008). However, *in vivo*, the efficacy of neurotrophic factors might vary greatly due to the method used for their delivering and it is thus important to deliver them near the regenerating site, in the physiologic concentrations. In this view, association of a cellular system producing neurotrophic factors to biodegradable membranes may greatly improve the nerve regeneration in neurotmesis (Geuna et al., 2009). N1E-115 cell line established from a mouse neuroblastoma (Kerns et al., 1991), might be a useful cellular system to locally produce and deliver these neurotrophic factors (Geuna et al., 2001). *In vitro*, the N1E-115 cells undergo neuronal differentiation in response to dimethylsulfoxide (DMSO), adenosine 3', 5'- cyclic monophosphate (cAMP), or serum withdrawal (Koka & Hadlock, 2001; Luis et al., 2007b; Luis et al., 2008; Luis et al., 2008). Upon induction of differentiation, proliferation of N1E-115 cells ceases, extensive neurite outgrowth is observed and the membranes become highly excitable (Koka & Hadlock, 2001; Luis et al., 2007b; Luis et al., 2008). The ideal interval period of 48 hours of differentiation was determined by measurement of the intracellular calcium concentration ($[Ca^{2+}]_i$), when the N1E-115 cells presented already the morphological characteristics of neuronal cells but at a time, cell death due to increased $[Ca^{2+}]_i$ was not still occurring (Rodrigues et al., 2005).

2. Biomaterial for tube-guides fabrication

The tube-guides or scaffold of tissue engineered nerve grafts serve in neurotmesis injuries to direct axons sprouting from the proximal to the distal stump, to maintain adequate mechanical support for the regenerating nerve fibers. Yet, they provide a conduit channel for the diffusion of neurotrophic and neurotropic factors secreted in the regeneration local and provide a conduit wall for the exchange of nutrients and waste products with the surrounding media, limiting the infiltration of fibrous scar tissue that hinders axonal regeneration (Johnson & Soucacos, 2008;). The neural scaffolds should be easy to be fashioned in different lengths and diameters, to be sterilized and implanted using microsurgical techniques. The neural scaffold has to satisfy many biological and physicochemical requirements like biocompatibility, biodegradability, permeability to ions, metabolites and for the revascularization of the regenerated nerve, biomechanical properties and surface properties that modulate the cellular system adhesion (Huang & Huang, 2006). A wide variety of biomaterials has been attempted which are of either natural or synthetic origin.

2.1 Natural biomaterials

The natural biomaterials are known by stimulating cell adhesion, migration, growth and proliferation. The natural biomaterials have also a very good biocompatibility and less toxic effects. Our research group has tested the collagen and the chitosan.

Collagen, laminin and fibronectin play an important role in the development and growth of axons (Grimpe & Silver, 2002). So, these components have become very important candidates for neural scaffolds, these materials can also serve as delivery vehicles for support cells, growth factors and drugs. For example, silicone tubes filled with laminin, fibronectin, and collagen led to a better regeneration over a 10 mm rat sciatic nerve gap compared to empty silicone controls (Chen et al., 2007). Collagen filaments have also been used to guide regenerating axons across 20–30 mm defects in rats (Itoh et al., 2003; Yoshii & Oka, 2001). Further studies have shown that oriented fibers of collagen within gels, aligned using magnetic fields, provide an improved template for neurite extension compared to randomly oriented collagen fibers (Ceballos et al., 1999). Finally, rates of regeneration comparable to those using a nerve autograft have been achieved using collagen tubes containing a porous collagen-glycosaminoglycan matrix (Archibald et al., 1995; Chamberlain et al., 2000). In our studies we have used equine collagen type III membrane (GentaFleece®; Baxter, Nuremberg, Germany) in axonotmesis and neurotmesis lesion of the rat sciatic nerve with very promising results (Amado et al., 2010; Luis et al., 2008).

Chitin is the second most abundant polysaccharide found in nature next to cellulose. It is a biopolymer of N-acetyl-D-glucosamine monomeric units and it has been used in a wide range of biomedical devices. Chitosan is a copolymer of D-glucosamine and N-acetyl-D-glucosamine and it has a very similar molecular structure with lamin, fibronectin and collagen. So, the chitosan has favorable biological properties for the nerve regeneration and it is easier to process than the chitin. Chitosan is quite fragile in its dry form, so, it has to undergo chemical cross-linking or to be used with other biomaterials before scaffold fabrication. The chitosan (high molecular weight, Aldrich®, USA) tested by our research group was dissolved in 0.25M acetic acid aqueous solution to a concentration of 2% (w/v). To obtain type III membranes, GPTMS (Aldrich®, USA) was also added to the chitosan solution and stirred at room temperature for 1h. The drying process for type III chitosan membrane was as follows: the solutions were frozen for 24h at -20°C and then transferred to the freeze-dryer, where they were left 12h to complete dryness. The chitosan type III membranes were soaked in 0.25N sodium hydroxide aqueous solution to neutralize remaining acetic acid, washed well with distilled water, and freeze dried (Amado et al., 2008). All membranes used in vivo testing, were sterilized with ethylene oxide gas, considered by some authors the most suitable method of sterilization for chitosan membranes (Marreco et al., 2004). Prior to their use *in vivo*, membranes were kept 1 week at room temperature to clear any ethylene oxide gas remnants (Amado et al., 2008; Simoes et al., 2010) (see Fig.1).

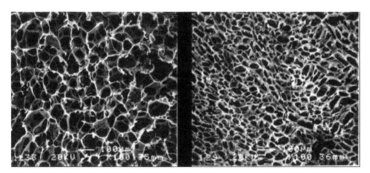

Fig. 1. SEM microstructure of chitosan membranes. Type II chitosan membrane (A). Type III chitosan membrane, showing a more porous microstructure, when compared to type II chitosan membrane (B) (Amado et al., 2008).

2.2 Synthetic biomaterials

The synthetic biopolymers constitute a group of promising biomaterials for the fabrication of neural tube-guides. However, the biocompatibility of some synthetic materials can be low since the difficulty of cell adhesion and survival can be a negative issue. Historically, non degradable synthetic materials were tested before the degradable ones. For instance, the silicone has been tested for peripheral nerve repair since 1960. Although it is not degradable in the body and it is impermeable to large molecules, the silicone tube-guide was an important model system for studying nerve regeneration. It was also applied in several clinical trials to bridge short nerve gaps with some success (Gu et al., 2011). The use of non-degradable tube-guides implies a second surgery time, which has clear disadvantages for the patient. In order to overcome the disadvantages associated with non-degradable materials, research has been focused in biodegradable synthetic biomaterials that should degrade within a reasonable time span. The degradation products absorbed should not be toxic or induce a foreign body reaction. Moreover, the physiochemical and biological properties of biodegradable synthetic materials can be tailored to match different application requirements, like to entrap some molecules or serve as a support for a cellular system. Aliphatic polyesters represent a class of the common degradable synthetic polymers, among which poly(L-lactic acid) (PLLA) (Wang et al., 2009), polyglycolic acid (PGA) (Waitayawinyu et al., 2007), polycaprolactone (PCL) (Mligiliche et al., 2003) and their copolymers, including poly(lactic-ε-caprolactone (Neurolac®) (Den Dunnen et al., 1998; Luis et al., 2007b) and poly(L-lactic-co-glycolic acid (PLGA) (Bini et al., 2004; Luis et al., 2007b; Luis et al., 2008). Among synthetic biodegradable tubes, two types attracted particular attention: those made of PLGA and those made of Poly (DL-lactide-ε-caprolactone) copolyester (Neurolac®) (see Fig.2).

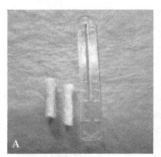 

Fig. 2. PLGA 90:10 (A) and poly (DL-lactide-e-caprolactona) copolyester (Neurolac®) tube-guides (B) (Luis et al., 2007b).

These two types of nerve guides showed several different physical-chemical properties. The Poly (DL-lactide-ε-caprolactone) tube-guides (Neurolac®), 16 mm long, internal diameter of 2 mm and thickness wall of 1.5 mm, were purchased from Polyganics BV, Groningen, The Netherlands (Luis et al., 2007b; Luis et al., 2008). Poly (dl-lactide-co-glycolide) copolymers with ratio of 90PLA:10PLG were obtained from their cyclic dimers, dl-lactide and glycolide. Non-woven constructs were used to prepare tube-guides 16 mm long, internal diameter of 2.0 mm and thickness wall of 1.5mm, to be applied in a 10-mm sciatic nerve gap. These fully synthetic non-woven materials are extremely flexible, biologically safe and are able to sustain the compressive forces due to body movement after implantation. They have also some degree of porosity to allow for influx of low molecular nutrients required for nerve regeneration. The non-woven structure allowed the tube-guide to hold suture without difficulties, however greater care had to be taken in order to ensure its integrity. These tube-guides of PLGA are expected to degrade to lactic and glycolic acids through hydrolysis of the ester bonds (Luis et al., 2007b; Luis et al., 2008).

The polyvinyl alcohol hydrogel (PVA) tube-guides are now starting to be tested by our research groups. Previous studies showed the PVA potential to several biomedical applications (Grant et al., 2006). The PVA gathers a set of exceptional properties, based in a nanofibrilar 3D structure, namely its high biocompatibility: it practically does not induce any foreign body or inflammatory reaction; it has high capacity to absorb water, mechanical resistance, elasticity, mobility and suturability (Bichara et al., 2010). The PVA are prepared in a tubular form, with 16 mm long, an internal diameter of 2 mm and thickness wall of 1.5 mm. These forms are prepared by a casting route to a mould that allows the final shape and thickness. Among the several methods described for the preparation of PVA material, the physical cross-linking of PVA by cyclic freezing and thawing is the one with better mechanical results (data not published by our research group). PVA from Aldrich (Mowiol 10-98: Mw 61,000, 98.0-98.8 mol% hydrolysis) is used. The tube-guides are produced using a solution of 20%(m/v) of PVA submitted to 3 repetitions of a freeze/thaw cycle, using 8 hours of freezing at –20.0°C and 4 hours of thawing at 22°C. However, several freeze/thaw cycling schemes must be tested to evaluate their effect on the final physical and mechanical properties, and choose the best one. The possible involvement of a final annealing step in the production process is being considered. This type of nerve guides can have unlimited availability in terms of diameters and lengths.

3. Cellular systems

The use of neural scaffolds alone to repair peripheral nerve defects have achieved variable success, but only small gap nerve repair (of 10 mm in rat sciatic nerve and of 30 mm in primate ulnar nerve) demonstrated evident functional and morphologic recovery (Amado et al., 2010; Gu et al., 2011; Hood et al., 2009; Luis et al., 2007b; Luis et al., 2008; Simoes et al., 2010). When the length of nerve gaps are increased, neural scaffolds alone only permit the bridging between the stumps, but for an effective regeneration, supporting cells or growth factors should be incorporated. The cellular systems implanted into the injury nerve may produce growth factors or ECM molecules, or may modulate the inflammatory process, to improve nerve regeneration (Amado et al., 2010; Gu et al., 2011; Luis et al., 2007b; Luis et al., 2008; Simoes et al., 2010). Schwann cells, mesenchymal stem cells, embryonic stem cells, marrow stromal cells are the most studied candidates of support cells among others. We focused our research in N1E-115 cell line *in vitro* differentiated (Amado et al., 2010; Luis et al., 2007b; Luis et al., 2008; Simoes et al., 2010) and in mesenchymal stem cells from Wharton's jelly of the umbilical cord (data not published yet).

To implant cultured cells (N1E-115 cells, MSCs, Schwann cells, and other cellular systems) into defective nerves (with axonotmesis and neurotmesis injuries), there are two main techniques. The cellular system may be directly injected to the neural scaffold which has been interposed between the proximal and distal nerve stumps or around the crush injury (in neurotmesis and axonotmesis injuries, respectively). In alternative, implant can also be achieved by pre-adding the cells to the neural scaffold via injection or co-culture (in most of the cellular systems, it is allowed to form a monolayer) and then the biomaterial with the cellular system is implanted in the injured nerve (Amado et al., 2010; Luis et al., 2008; Luis et al., 2008; Simoes et al., 2010).

3.1 Cell culture of N1E-115 cells – a cellular model for stem cells transplantation

The role of neurotrophic factors in neural regeneration has been the focus of extensive research in nerve regeneration (Schmidt & Leach, 2003). The influence of these factors in neural development, survival, outgrowth, and branching has been explored on various levels, from molecular interactions to macroscopic tissue responses. One family of neurotrophic factors, the neurotrophins, has been heavily investigated in nerve regeneration studies, including the nerve growth factor (NGF), brain-derived neurotrophic factor (BDNF), neurotrophin-3 (NT-3), and neurotrophin-4/5 (NT-4/5) (Lundborg et al., 1994). Neurotrophic factors promote a variety of neural responses: survival and outgrowth of the motor and sensory nerve fibers, spinal cord and peripheral nerve regeneration (Schmidt & Leach, 2003). However, *in vivo* responses to neurotrophic factors can vary due to the method of their delivering. Therefore, the development and use of highly controlled delivery devices are required for the study of these extremely complex systems. For that reason, N1E-115 cell line established from neuroblastoma, that undergo neuronal differentiation in response to dimethylsul-foxide (DMSO), adenosine 3′;5′-cyclic monophosphate (cAMP), or serum withdrawal (Amado et al., 2010; Rodrigues et al., 2005) was a cellular system tested in axonotmesis and neurotmesis lesions to locally produce and deliver these neurotrophic factors. Upon induction of differentiation, proliferation of N1E-115 cells ceases, extensive neurite outgrowth is observed and the membranes become highly excitable (Amado et al., 2010; Rodrigues et al., 2005). During differentiation, cyclin-dependent kinase (cdk) activities decline and phosphorylation of the retinoblastoma gene product (pRb) is lost, leading to the appearance of a pRb-containing E2F DNA-binding complex. The molecular mechanism by which pRb inhibits cell proliferation is becoming increasingly clear. pRb inhibits the activity of proteins that function as inducers of DNA synthesis (Kranenburg et al., 1995). Ca^{2+} serves as an important intracellular signal for cellular processes such as growth and differentiation. Free calcium levels within neural cells control many essential neural functions including neurotransmitter release, membrane conductance, nerve fibber excitability, coupling between neuronal cells, and axonal transport. Although regulation of the intracellular Ca^{2+} concentration ($[Ca^{2+}]_i$) is important for normal cell functioning, its deregulation has been linked to cellular pathologies and cell death (Trump & Berezesky, 1995). Deregulation in $[Ca^{2+}]_i$ can be toxic to cells and is involved in the triggering of events leading to excitotoxic cell death in neurons, through the activation of calpain, phospholipases and endonucleases (Smith & Hall, 1988). Because of the importance of $[Ca^{2+}]_i$ in neuronal health and disease, a relatively simple cell model system, where $[Ca^{2+}]_i$ regulation can be studied fairly easily, is desirable. We have tested a non-expensive and easy method to culture neural-like cell line capable of producing locally, nerve growth factors and of growing inside PLGA90:10, chitosan and collagen tubular/membrane nerve guides (Amado et al., 2010; Amado et al., 2008; Luis et al., 2008; Luis et al., 2008; Simoes et al., 2010), in order to use them to promote nerve regeneration across a peripheral nerve gap. To correlate the neuronal cells' ability to promote nerve regeneration across a gap, through their differentiation grade and survival capacity, the $[Ca^{2+}]_i$ of non-differentiated and DMSO differentiated N1E-115 cells was determined by the epifluorescence technique using the Fura-2-AM probe (Tsien, 1989). The measurement of $[Ca^{2+}]_i$ permitted to determine an ideal period of differentiation of 48 hours, when the N1E-115 cells presented already the morphological characteristics of neuronal cells and at the same time, the death process was not initiated by the $[Ca^{2+}]_i$ modifications. This cellular system was used in axonotmesis and neurotmesis injuries associated to chitosan, equine collagen type III and PLGA membranes or tube-guides in order to improve the functional and morphologic recovery of the nerve (Amado et al., 2010; Gu et al., 2011; Luis et al., 2007b; Luis et al., 2008; Simoes et al., 2010).

3.2 Wharton's jelly mesenchymal cells

After peripheral nerve injury, many neurons die. Because neurons cannot be easily expanded in vitro, there is a wide difficulty in applying primary cultured neurons to nerve tissue engineering. Embryonic stem cells (ESCs) have a great potential to proliferate unlimitedly and differentiate into neural cells using several differentiation protocols (Gu et al., 2011). The differentiation of ESCs can be modulated by growth factors and retinoic acid (Gu et al., 2011). Extensive research has been focused on the implantation of these stem cells in Central Nervous System (CNS) disorders, but in Peripheral Nerve System (PNS) injuries only a few studies were published. Cui and co-workers, in 2008, implanted ESC-derived neural progenitor cells into a 10-nm sciatic nerve gap, resulting in substantial axonal regrowth and nerve repair (Cui et al., 2008). The implanted cells survived for 3 months and could differentiate into myelinating cells. The nerve stumps presented almost normal diameter with longitudinally oriented, densely packed Schwann cell-like arrangement. The regenerated nerves also showed recovery of the functional activity, determined by electrophysiological records (Cui et al., 2008). Although these promising results, the ESCs have serious ethical issues, especially when obtained by somatic nuclear transfer technique or from embryos produced for assisted medical reproduction.

Recent years have witnessed a great expansion in the number of adult stem cell populations isolated and characterized. Adult stem cells have long been considered restricted, considering their multipotency. Many studies have challenged this dogma, suggesting that adult stem cells may be more plastic than previously appreciated (Beer et al., 2001). Extra-embryonic tissues as stem cell reservoirs offer many advantages over embryonic and adult stem cell sources, and are routinely discarded at parturition. Little ethical controversy attends the harvest of the resident stem cell populations. Human umbilical cord (UC) consists of three tissue components including the (1) amniotic membrane, (2) stroma (namely, Wharton's jelly), and (3) blood vessels (two arteries and one vein). The stroma is also further divided into three histological zones including the (1) subamniotic zone, (2) intervascular zone, and (3) perivascular zone (Can & Karahuseyinoglu, 2007). Wharton's jelly (WJ) is a proteoglycan rich connective tissue of the UC. Within the abundant extracellular matrix of WJ resides a recently described stem cell population called UC Matrix Stem Cells. Can be isolated around 400,000 cells per UC, which is significantly greater than the number of MSCs that can be routinely isolated from adult bone marrow. Isolated MSCs express CD29 and CD54, consistent with a mesenchymal cell type, and can be propagated in culture for more than 80 population doublings (Marcus & Woodbury, 2008). In vitro the WJ MSCs cells are capable of differentiation to multiple mesodermal cell types. This cells have been successfully differentiated into various cell types including adipocytes (Karahuseyinoglu et al., 2008), chondrocytes (Baksh et al., 2007; Wang et al., 2004), osteocytes (Baksh et al., 2007; Conconi et al., 2006; Wang et al., 2004), cardiomyocytes (Kadivar et al., 2006; Wang et al., 2004), skeletal myocytes (Conconi et al., 2006), hepatocytes (Kadivar et al., 2006), insulin-producing cells (Wu et al., 2009) as well as neuron-like cells (Fu et al., 2006; Weiss et al., 2006) (see Fig.3).

Generation of clinically important dopaminergic neurons has been also reported by other groups (Fu et al., 2006; Wang et al., 2004). MSCs isolated from WJ may be a valuable cellular system to improve peripheral nerve regeneration, since they possess stem cell properties and are able to differentiate in vitro into neuron-like cells (Fu et al., 2006; Yang et al., 2008). The transplanted cells are able to promote local blood vessel formation and to locally produce neurotrophic factors like BDNF and GDNF (Wang et al., 2004). Transformed MSCs are still viable 4 months after transplantation without the need for immunological suppression, suggesting them as good source for Regenerative Medicine (Fu et al., 2006).

 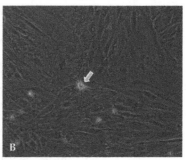

Fig. 3. Undifferentiated mesenchymal stem cells, from Wharton's jelly, exhibiting a star-like shape with a flat morphology (A) and neural-like cells, differentiated after 96 hours with neurogenic medium, formation of typical neural-like multi-branches (B) (100X magnification).

4. Microsurgical procedures

Rodents, particularly the rat and the mouse, have become the most frequently utilized animal models for the study of peripheral nerve regeneration because of the widespread availability of these animals as well as the distribution of their nerve trunks which is similar to humans (Mackinnon et al., 1985; Rodriguez & Navarro, 2004). Although other nerve trunks, especially in the rat forelimb, are getting more and more used for experimental research (Papalia et al., 2006), the rat sciatic nerve is still the far more employed experimental model as it provides a nerve trunk with adequate length and space at the mid-thigh for surgical manipulation and/or introduction of grafts or guides (Rodriguez & Navarro, 2004). Although sciatic nerve injuries themselves are rare in humans, this experimental model provides a very realistic testing bench for lesions involving plurifascicular mixed nerves with axons of different size and type competing to reach and reinnervate distal targets (Mackinnon et al., 1985). Common types of experimentally induced injuries include focal crush or freeze injury that causes axonal interruption but preserves the connective sheaths (axonotmesis), complete transection disrupting the whole nerve trunk (neurotmesis) and resection of a nerve segment inducing a gap of certain length. For these reasons our experimental work, concerning the *in vivo* testing of neurotmesis injury regeneration, was based on the use of Sasco Sprague-Dawley rat sciatic nerve. Usually, the surgeries are performed under an M-650 operating microscope (Leica Microsystems, Wetzlar, Germany). Under deep anaesthesia (ketamine 9 mg/100 g; xylazine 1.25 mg/100 g, atropine 0.025 mg/100 body weight, intramuscular), the right sciatic nerve is exposed through a skin incision extending from the greater trochanter to the distal mid-half followed by a muscle splitting incision. After nerve mobilisation, a transection injury is performed (neurotmesis) using straight microsurgical scissors, at a level as low as possible, in general, immediately above the terminal nerve ramification. The proximal and distal nerve stumps are inserted 3 mm into the tube-guide (not covered or covered with cells) and held in place, maintaining a nerve gap of 10mm, with two epineurial sutures using 7/0 monofilament nylon. For neurotmesis without gap, the nerve is reconstructed with an end-to-end suture, with two epineural sutures using de 7/0 monofilament nylon. Finally the skin and subcutaneous tissues are closed with a simple-interrupted suture of a non-absorbable filament (Synthofil®, Ethicon). An antibiotic (enrofloxacin, Alsir® 2.5 %, 5 mg / kg b.w., subcutaneously) is always administered to prevent any infections. To prevent autotomy a deterrent substance must be applied to rats' right foot (Kerns et al., 1991; Sporel-Ozakat et al., 1991). All procedures must be performed accordance with the European Communities Council Directive of November 1986 (86/609/EEC).

5. Functional evaluation

Our studies of sciatic nerve regeneration after neurotmesis include a post-surgery follow-up period of 20 weeks based on the assumption that, by the end of this time, functional and morphological recovery are complete (Amado et al., 2010; Luis et al., 2007b; Luis et al., 2008; Simoes et al., 2010). Although both morphological and functional data have been used to assess neural regeneration after induced neurotmesis and axonotmesis injuries, the correlation between these two types of assessment is usually poor (Dellon & Mackinnon, 1989; Shen & Zhu, 1995). Classical and newly developed methods of assessing nerve recovery, including histomorphometry, retrograde transport of horseradish peroxidase and retrograde fluorescent labelling (Mackinnon et al., 1985; Mackinnon SE, 1988) do not necessarily predict the reestablishment of motor and sensory functions (de Medinaceli et al., 1982; Shen & Zhu, 1995). Although such techniques are useful in studying the nerve regeneration process, they generally fail in assessing functional recovery (Shen & Zhu, 1995). In this sense, research on peripheral nerve injury needs to combine both functional and morphological assessment. The use of biomechanical techniques and rat's gait kinematic evaluation is a progress in documenting functional recovery (Varejao et al., 2003b). Indeed, the use of biomechanical parameters has given valuable insight into the effects of the sciatic denervation/reinnervation, and thus represents an integration of the neural control acting on the ankle and foot muscles (Varejao et al., 2003b; Varejao et al., 2002).

5.1 Evaluation of motor performance (EPT) and nociceptive function (WRL)
Motor performance and nociceptive function were evaluated by measuring extensor postural thrust (EPT) and withdrawal reflex latency (WRL), respectively. The animals are tested pre-operatively (week-0), at weeks 1, 2, and every two weeks thereafter until week-20. The EPT was originally proposed by Thalhammer and collaborators, (Thalhammer et al., 1995) as a part of the neurological recovery evaluation in the rat after sciatic nerve injury. For this test, the entire body of the rat, excepting the hind-limbs, is wrapped in a surgical towel. Supporting the animal by the thorax and lowering the affected hind-limb towards the platform of a digital balance, elicits the EPT. As the animal is lowered to the platform, it extends the hind-limb, anticipating the contact made by the distal metatarsus and digits. The force in grams (g) applied to the digital platform balance is recorded. The same procedure is applied to the contra-lateral, unaffected limb. Each EPT test must be repeated 3 times and the average result is considered. The normal (unaffected limb) EPT (NEPT) and experimental EPT (EEPT) values are incorporated into a equation (Equation 1) to derive the percentage of functional deficit (Koka & Hadlock, 2001).

$$\text{Percentage motor deficit} = [(NEPT - EEPT) / NEPT]*100$$

The EPT data, originally measured in grams of weight borne by each hindlimb, is expressed as percentage deficit from total bearing weight as determined by the weight borne by the unoperated limb. Potential pitfalls of the EPT do exist. It takes a certain training period for the tester to become comfortable handling the animals, and this comfort level is critical for the animal to behave in a non frightened way. There is also a level of recognition of when the animal is bearing its maximum weight, which is critical since the tester is supporting the body of the animal at all times. Once this recognition takes place, through training by an experienced tester, the examination becomes highly reproducible (Koka & Hadlock, 2001).

To assess the nociceptive withdrawal reflex (WRL), the hotplate test was modified as described by Masters and collaborators (1993). The rat is wrapped in a surgical towel above its waist and then positioned to stand with the affected hind paw on a hot plate at 56°C and with the other on a room temperature plate. WRL is defined as the time elapsed from the onset of hotplate contact to withdrawal of the hind paw and measured with a stopwatch. Normal rats withdraw their paws from the hotplate within 4.3 s or less (Hu et al., 1997). The affected limbs are tested 3 times, with an interval of 2 min between consecutive tests to prevent sensitization, and the three latencies are averaged to obtain a final result (Campbell, 2001). If there is no paw withdrawal after 12 s, the heat stimulus is removed to prevent tissue damage, and the animal is assigned with the maximal WRL of 12 s (Varejao et al., 2003a).

5.2 Sciatic functional index (SFI) and static sciatic index (SSI)

For SFI, animals are tested in a confined walkway measuring 42-cm-long and 8.2-cm-wide, with a dark shelter at the end. A white paper is placed on the floor of the rat walking corridor. The hind paws of the rats are pressed down onto a finger paint-soaked sponge, and they are then allowed to walk down the corridor leaving its hind footprints on the paper. Often, several walks are required to obtain clear print marks of both feet. Prior to any surgical procedure, all rats are trained to walk in the corridor, and a baseline walking track is recorded. Subsequently, walking tracks are recorded pre-operatively (week-0), at weeks 1, 2, and every two weeks thereafter until week-20. Several measurements are taken from the footprints: (I) distance from the heel to the third toe, the print length (PL); (II) distance from the first to the fifth toe, the toe spread (TS); and (III) distance from the second to the fourth toe, the intermediary toe spread (ITS). The static footprints are obtained at least during four occasional rest periods. In the static evaluation (SSI) only the parameters TS and ITS, are measured. For both dynamic (SFI) and static assessment (SSI), all measurements are taken from the experimental and normal sides. Four steps should be analyzed per rat. Prints for measurements should be chosen at the time of walking based on clarity and completeness at a point when the rat was walking briskly. The mean distances of three measurements are used to calculate the following factors (dynamic and static):

$$\text{Toe spread factor (TSF)} = (ETS - NTS)/NTS$$

$$\text{Intermediate toe spread factor (ITSF)} = (EITS - NITS)/NITS$$

$$\text{Print length factor (PLF)} = (EPL - NPL)/NPL$$

Where the capital letters E and N indicate injured and non-injured side, respectively.

The SFI was calculated as described by Bain et al. (1989) (Bain et al., 1989) according to the following equation:

$$SFI = -38.3 \ (EPL - NPL)/NPL + 109.5(ETS\text{-}NTS)/NTS + 13.3(EIT\text{-}NIT)/NIT - 8.8 = (-38.3 \times$$
$$PLF) + (109.5 \times TSF) + (13.3 \times ITSF) - 8.8$$

The SSI is a time-saving and easy technique for accurate functional assessment of peripheral nerve regeneration in rats and is calculated using the static factors, not considering the print length factor (PL), according to the equation (Bervar, 2000; Meek et al., 2004)

$$SSI = [(108.44 \times TSF) + (31.85 \times ITSF)] - 5.49$$

For both SFI and SSI, an index score of 0 is considered normal and an index of -100 indicates total impairment. When no footprints are measurable, the index score of -100 is given. In each walking track three footprints must be analyzed by a single observer, and the average of the measurements are used in SFI calculations.

5.3 Kinematic analysis

Locomotion is also of higher functional relevance since it involves integrated function of both the motor and sensory systems and their respective components, such as skeletal muscles, sensory endings, efferent and afferent nerve fibers and integrative centers within the central nervous system. Muscles innervated by sciatic nerve branches include both dorsiflexors and plantarflexors and, although in previous studies we focused our kinematic analysis only in the stance phase (Amado et al., 2010; Amado et al., 2008; Luis et al., 2008), we now prefer to include analysis of the ankle joint motion also during the swing phase in order to provide additional information (Joao et al., 2010).

In our previous studies we analyzed ankle kinematics in the sagittal plane during the stance phase of walking either after sciatic nerve transection and repair (Amado et al., 2010; Luis et al., 2008) or after sciatic nerve crush (Amado et al., 2008). The two-dimensional (2D) ankle motion analysis was conducted during voluntary level walking along a corridor with darkened cages at both ends to attract the animals. The lateral walls of the corridor were made of Perspex and a high speed video camera (JVC GR-DVL9800, Nex Jersey, USA) was placed orthogonally to corridor length in order to record ankle motion during walking. Sagittal records of the rat walking were obtained at a frame rate of 100 frames per second and images were semi-automatically digitized using marks at reference points of the rat hindlimb and foot (see Fig. 4B).

 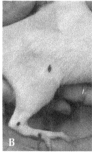

Fig. 4. Reflective markers with 2mm diameter were attached to the right hindlimb at bony prominences (from bottom to top): (1) tip of fourth finger, (2) head of fifth metatarsal, (3) lateral malleolus, (4) lateral knee joint, (5) trochanter major (A). Tattoo marks at bony prominences of the right hindlimb: (1) head of fifth metatarsal, (2) lateral malleolus, (3) lateral knee joint (B).

The trajectories of the segments leg and hindfoot were obtained through this procedure and ankle joint angle was derived by using dot product computation. Ankle kinematics parameters proposed by Varejão et al. (2003) were employed in order to assess functional recovery after sciatic nerve transection and repair and to compare treatments using our tested biomaterials and cellular system. Our ankle joint kinematic analysis generally showed profound walking dysfunction in the weeks after sciatic neurotmesis. Changes affected the whole stance phase and were best characterized by an inability of the animals to perform the normal ankle push off action during the second half of this walking phase (see Fig 5 and Fig.6). Profound ankle kinematic changes during the stance phase of walking were also evident when taking into account angular velocity data. Uninjured animals show a clear ankle angular velocity peak both at the initiation and end of the stance phase. After sciatic transection these two peaks are not observed and the stance phase is characterized by a rather steady increase in positive   ankle velocity that corresponds to a progressive augment in ankle dorsiflexion. This abnormal pattern of ankle motion during the stance phase of walking was interpreted as caused by denervation and paralysis of the ankle plantarflexors that were unable to actively extend the ankle against the load of the body weight. In sciatic nerve transected and repaired animals, the recovery of the normal ankle motion pattern of walking is slow and largely incomplete. After 20 weeks of recovery, abnormal ankle motion is still noticed after sciatic nerve neurotmesis (Amado et al., 2010; Luis et al., 2008). In our recent study (Amado et al., 2010), investigating the role of a collagen membrane with and without differentiated N1E-115 cells enwrapped around the transected and end-to-end repaired sciatic nerve, the shape of ankle joint motion during the stance phase showed little improvements until three months after injury (see Fig 5 and Fig.6).

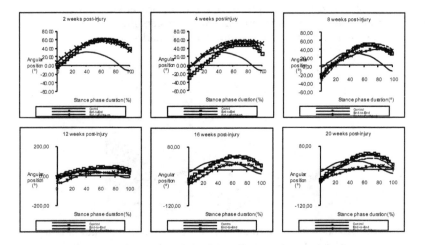

Fig. 5. Kinematics plots in the sagittal plane for the angular position (°) of the ankle as it moves through the stance phase, during the healing period of 20 weeks. The mean of each group is plotted.

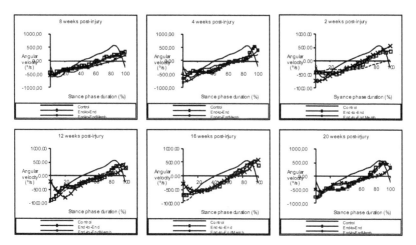

Fig. 6. Kinematics plots in the sagittal plane for the angular velocity (°/s) of the ankle as it moves through the stance phase, during the healing period of 20 weeks. The mean of each group is plotted.

A progressive improvement in ankle kinematics occurred from this time point up to the end of the 20 weeks recovery time. Such improvement consisted mainly in a slight recovery of ankle plantarflexion at the end of stance phase and lesser peak dorsiflexion at midstance. These changes in ankle kinematics are suggestive of muscle reinnervation and increased force generation ability by the ankle pantarflexor muscle group. In this study, ankle kinematics data were in general agreement with the results of the EPT and WRL tests. The latter tests also demonstrated progressive recovery of motor and sensory function along the follow up time that in the case of the WRL test was accelerated beyond the 12 weeks time point. As for ankle kinematics, the EPT and WRL results also demonstrated incomplete functional recovery in these animals, with slightly better recovery in the end-to-end repair group compared to other two groups. However, ankle kinematics must be envisaged merely as an indirect measure of muscle function. It should be noticed that walking requires fine coordination of limbs motion and definitely quadruped animals have many movement strategies to compensate for deficit in one of the hindlimbs. Through plasticity of integrative structures, animals may develop adaptive patterns that persist even if significant reinnervation takes place. Also, no direct relation exists between more simple tests of muscle strength or of sensory function and a complex action such as walking (Varejao et al., 2003b). Indeed, ankle kinematics data poses a fundamental question. Does recovery of ankle motion during walking signals successful muscle reinnervation and regain of muscle function or else is a product of compensatory changes at whole hindlimb level? To answer this question satisfactorily, walking analysis has to be improved by extending the analysis to hip and knee joints and should be combined with direct measures of muscle activity (Gramsbergen et al., 2000a) and to measures of the force applied to the ground by the walking animal (Howard et al., 2000). To better assess hindlimb joint kinematics during walking, we recently analyzed hip, knee and ankle joint kinematics during recovery of less severe sciatic nerve crush injury, using a more sophisticated motion capture system to track the motion of reflective markers attached to the rat hindlimb (unpublished observations) (see Fig. 4A).

After this kind of injury, functional recovery is often deemed as complete even on the basis of ankle kinematic analysis with only minor walking changes observed after 12 weeks of recovery. However, after 12 weeks of recovery from complete sciatic nerve crush, changes in hip and knee kinematics during walking were present, when compared to sham-operated control animals. Importantly, such altered hip and knee kinematics were in contrast with total reestablishment of the normal pattern of ankle motion in sciatic-crushed animals.

Individual joint kinematics either in control or nerve-injured animals is characterized by high variability, with notable differences between different animals and even from step to step (Chang et al., 2009). Such high level of variability, which seems to be an intrinsic property of normal quadruped walking, seriously affects the precision of joint kinematic measures of functional recovery after nerve injury. Reducing this variability is a challenge for efficient use of walking analysis to assess functional recovery. Attempts to overcome this limitation include constraining walking velocity by using treadmill walking instead of self-paced locomotion (Pereira et al., 2006). This, of course, is likely to reduce step-by-step variability in joint kinematics but has the disadvantage of requiring expensive equipment and limits the possibility of combining kinematic analysis with other data, such as ground reaction forces. Other possibilities look at a global, limb-level movement analyses as an alternative to individual joints kinematics (Chang et al., 2009; Sabatier et al., 2011). Also, systematic changes in the biomechanical and movement control constraints of the locomotor task, such as using up- and down-slope walking might also increase the accuracy of walking analysis within the context of peripheral nerve research (Sabatier et al., 2011).

In summary, walking analysis is a promising method to assess functional recovery after hindlimb nerve injury. However, in order to provide accurate measures of functional recovery, walking analysis after hindlimb peripheral nerve injury will have to evolve from simply analyzing ankle kinematics to reach a full biomechanical description of hindlimb motion including analysis of hip, knee and ankle joints. Further refinements of walking analysis in the field of peripheral nerve research using the rat model will probably include the combined use of joint kinematics, ground reaction forces and electromyographical data of muscle activity.

6. Morphological analysis

It has been recently pointed out that morphological analysis is the far most common method for the study of peripheral nerve regeneration (Raimondo et al., 2009). Actually, the investigation of nerve morphology can give us important information on various aspects of the regeneration processes which relates with nerve function (Geuna et al., 2009).

Although different types of fixatives can be used for peripheral nerve histology, we fix nerve samples in a solution of 2.5% purified glutaraldehyde (Histo-line Laboratories s.r.l., Milano, Italy) and 0.5% saccarose (Merck, Darmstadt, Germany) in 0.1M Sörensen phosphate buffer, pH 7.4, for 6-8h. Nerves are then washed and stored in 0.1M Sörensen phosphate buffer added with 1.5% saccarose at 4-6°C prior to embedding. Sörensen phosphate buffer is made with 56g di-potassium hydrogen phosphate 3-hydrate (K2HPO4-3H2O) (Fluka, Buchs, Switzerland) and 10.6 g sodium di-hydrogen phosphate 1-hydrate (NaH2PO4-H2O) (Merck, Darmstadt, Germany) in 1 litre of doubly-distilled water. Just before the embedding, nerves are washed for few minutes in the storage solution and then immersed for 2 h in 2% osmium tetroxide (Sigma, St.Louis, MO) in the same buffer solution. The specimens are then carefully dehydrated in passages in ethanol and embedded in

Glauerts' mixture of resins which is made of equal parts of Araldite M and the Araldite Härter, HY 964 (Merck, Darmstad, Germany). At the resin mixture, 2% of accelerator 964, DY 064 is added (Merck, Darmstad, Germany). Finally, the plasticizer (0.5% of dibutylphthalate) is added to the resin.

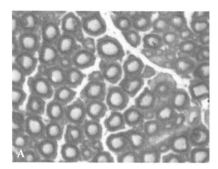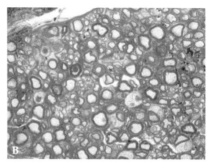

Fig. 7. Representative high resolution photomicrographs of nerve fibers from normal (A) and regenerated (B) rat sciatic nerves after entubulation with collagen nerve guide. Original magnification = 1,000x.

In our laboratory, histomorphometry (stereology) is carried out on toludine-blue-stained semi-thin sections (2.5 micron-thick) of nerve samples using a DM4000B microscope equipped with a DFC320 digital camera and an IM50 image manager system (Leica Microsystems, Wetzlar, Germany). We adopt a final magnification of 6,600x in order to enable accurate identification of myelinated nerve fibers (Figure 9). A 2D-disector method, (Raimondo et al., 2009) is finally used for estimating the total number of myelinated fibers (N), the mean diameter of fiber (D) and axon (d) as well as mean [(D-d)/2] and g-ratio (D/d) (see Fig.7).

7. Synopsis of results and discussion

Peripheral nerve regeneration can be studied in a number of different experimental models based on the use of nerves from both forelimb and hindlimb (Hu et al., 1997). The experimental animal model of choice for many researchers remains the rat sciatic nerve. It provides an inexpensive source of mammalian nervous tissue of identical genetic stock that it is easy to work with and well studied (Geuna et al., 2009) and shows a similar capacity for regeneration in rats and sub-humans primates (Johnson et al., 2008). The rat sciatic nerve is a widely used model for the evaluation of motor as well as sensory nerve function at the same time. One of the most addressed issues in experimental nerve repair research is represented by entubulation (Chen et al., 2006). While early studies were more directed towards biological material (Kerns et al., 1991) recent studies have focused more on synthetic materials that are biodegradable. The majority of natural biomaterials used in clinical applications such as hyaluronic acid, collagen, and gelatin are derived from animal sources. In spite of thorough purification methods these materials bear the inherent risk of transfer of viral and prionic diseases and may cause immunological body reactions (Koka & Hadlock, 2001; Luis et al., 2007b). Synthetic biomaterials are not associated with these risks (Koka & Hadlock, 2001; Luis et al., 2007b).

Our group has assessed two types of nerve guides made of a different type of biodegradable synthetic biomaterial, PLGA and caprolactone (Neurolac®). Our data showed that both types of nerve guides were a mean to help in the growth of axons and didn't deleteriously interfere with the nerve regeneration process. While the information on the effectiveness of Neurolac® tubes for allowing nerve regeneration was already provided experimentally (; Geuna et al., 2009; Luis et al., 2008) and with patients (Bertleff et al., 2005), the PLGA tubes with the polymers proportion of 90 PLA:10 PLG used in that study have never been tested *in vivo* before (Luis et al., 2007b). Results of the comparative functional assessment showed that the differences between the two experimental groups were not significant in terms of motor and nociceptive recovery. The EPT proposed by Thalhammer and coworkers (Geuna, 2005) to evaluate the motor function in rats is used routinely by veterinarian neurologists to evaluate the nervous system of small domestic animals. This reflex is initiated by a stretching of the spindles in the interosseous muscles and stimulation of sensory receptors of the foot (Lundborg, 2002). A steady recover of motor deficit occurred throughout the 20-week post-surgery period in the group of animals where PLGA and Neurolac® tube-guides were used. As expected, the motor recovery in animals of the end-to-end group occurred significantly faster and to a larger extent when compared to the entubulation groups. Similarly, nociception recovered to a significantly larger extent in the end-to-end group compared to both entubulation groups which, on the contrary, did not differ significantly between them (Luis et al., 2007b). Interestingly, when the nerve is transected and the regenerating axons must bridge a gap, sensory neurons exhibit a faster regenerative pattern than motor neurons (Lundborg, 2002). Morphological analysis showed a different pattern of biodegradation of the two tubes. In fact, while in the nerves repaired by Neurolac® the structure of the polymer was still well preserved at week-20, in the nerves repaired by PLGA guides, the polymer was largely substituted by a connective matrix rich in collagen fibbers and fibroblasts resembling a normal epineurium. These results may suggest that PLGA can be a better biomaterial for fashioning nerve guides since nerves regenerated inside it were more similar to a normal nerve than those regenerated inside Neurolac® guides, although the results of the motor and sensory functional assessment did not disclose significant differences between the two tube-guides and thus both should be considered a good substrate for preparing tubular nerve guides. We have also used the rat sciatic nerve model for investigating the effects of chitosan type III membranes after neurotmesis followed by surgical repair either by direct suture, or autograft or tubulization. Morphological results showed that nerve regeneration occurred when the chitosan type III tube-guide was used and that, at time of withdrawal, Wallerian degeneration was almost completed and substituted by re-growing axons and the accompanying Schwann cells. The results obtained with chitosan type III tube-guides were significantly better, in terms of functional and morphologic recovery, when compared with PLGA90:10 where the regeneration pattern was worse (Simoes et al., 2010). The synergistic effect of a more favorable porous microstructure and physicochemical properties (more wettable and higher water uptake level) of chitosan type III compared to common chitosan, as well as the presence of silica ions, may be responsible for the good results in promoting post-traumatic nerve regeneration (Amado et al., 2008) suggesting that this material may not just work as a simple mechanical scaffold but instead may work as an inducer of nerve regeneration (Amado et al., 2008). The neuroregenerative property of chitosan type III might be explained by a direct stimulation of Schwann cell proliferation, axon elongation and myelination (Shirosaki et al., 2005). Yet, the expression of established myelin genes such as PMP22, PO and MBP (Pietak et al., 2007) may be influenced by the presence of silica ions which exert an effect on several glycoprotein expression (Pietak et al., 2007). Taken together, results of this study support the view that hybrid chitosan type III membranes can be a valuable tool for fashioning nerve guides aimed to bridge nerve defects. We also investigated the effects on nerve regeneration of the employment PLGA90:10 nerve guide tubes covered with N1E-115 cells in vitro differentiated in the presence of DMSO for 48 hours (Geuna et al., 2001). Results showed that both motor and sensory functions improved significantly similarly to what occurs in PLGA tubes alone (Luis et al., 2008). Nerve recovery of around 60% was achieved by week-20, in both groups, reconstructed with PLGA 90:10 tube-guides with or without the cellular system. The pattern of degradation and the degradation products of PLGA90:10 more acidic than the collagen ones do not seem to influence negatively the degree of nerve regeneration during the healing period. On the other hand, the biodegradable non-woven structure of PLGA tubes may offer some advantages in terms of microporosity which may enhance nerve repair. It is not surprising that recovery was significantly better in the group where the gap was reconstructed using the autologous graft since this is still considered the gold standard for peripheral nerve regeneration (Matsuyama et al., 2000). The rationale for the use of the N1E-115 cells was to take advantages of the properties of these cells as a neural-like cellular source of neurotrophic factors (Geuna et al., 2001).

Although tube guides are a suitable choice for peripheral nerve reconstruction, it is known that nerve regeneration and functional recovery are less satisfactory than when nerve repair is done by using an end-to-end neurorraphy or when an autologous nerve graft is applied (Geuna et al., 2009). For example, the risk of neuroma formation in neurotmesis injury is considerable, which is clearly avoided with tube-guide technique. In this sense, enhancing the rate of axonal growth might prevent the occurrence of side effects and might turn nerve regeneration faster, thus improving functional recovery. Neurotrophic factors play an important role in nerve regeneration after injury or disease and it is conceivable that if neurotrophic factors are applied in the close vicinity of the injured nerve their healing potency is optimized. In spite of these premises and contrary to our initial hypothesis, the N1E-115 cells did not facilitate either nerve regeneration or functional recovery and, as far as morphometrical parameters are concerned, results showed that the presence of this cellular system reduced the number and size of the regenerated fibers. These results suggest that this type of nerve guides can partially impair nerve regeneration, at least from a morphological point of view (Luis et al., 2008). The impaired axonal regeneration seems to be the result of N1E-115 cells surrounding and invading the regenerating nerve, since numerous of these cells where seen colonizing the nerve and might have deprived regenerated nerve fibers blood supply (Luis et al., 2008). The use of N1E-115 cells did not promote nerve healing and their use might even derange the nerve regenerating process. Whereas the effects on nerve regeneration were negative, an interesting result of this study was the demonstration that the cell delivery system that we have used was effecting in enabling long-term colonization of the regenerated nerve by transplanted neural cells. Whether the negative effects of using N1E-115 cells as a cellular aid to peripheral neural tissue regeneration extends to other types of cells is not known at present and further studies are warranted to assess the role of other cellular systems, e.g. mesenchymal stem cells, as a foreseeable therapeutic strategy in peripheral nerve regeneration. Our experimental results with this cellular system are also important in the perspective of stem cell transplantation employment for improving posttraumatic nerve regeneration with patients (Luis et al., 2008). Undoubtedly, great enthusiasm has raised among researchers and in the general public about cell-based therapies in regenerative medicine (Geuna et al., 2009; Thalhammer et al., 1995). There is a widespread opinion that this type of therapy is also very safe in comparison to other pharmacological or surgical therapeutic approaches. By contrast, recent studies showed that cell-based therapy might be ineffective for improving nerve regeneration (Varejao et al., 2003b) or even have negative effects by hindering the nerve regeneration process after tubulisation repair. Whereas the choice of the cell type to be used for transplantation is certainly very important for the therapeutic success, our present results suggest that the paradigm that donor tissues guide transplanted stem cells to differentiate in the direction that is useful for the regeneration process it is not always true and the possibility that transplanted stem cells choose another differentiation line potentially in contrast with the regenerative process should be always taken in consideration.

8. Acknowledgements

This work was supported by the Fundação para a Ciência e Tecnologia (FCT) through the PhD grant SFRH/BD/70211/2010. Financial support was also provided by the European Regional Development Fund, through the QREN Project BIOMAT&CELL No. 1372.

9. References

Amado, S., Rodrigues, J. M., Luis, A. L., Armada-da-Silva, P. A., Vieira, M., Gartner, A., Simoes, M. J., Veloso, A. P., Fornaro, M., Raimondo, S., Varejao, A. S., Geuna, S., & Mauricio, A. C. (2010). Effects of collagen membranes enriched with in vitro-differentiated N1E-115 cells on rat sciatic nerve regeneration after end-to-end repair. *J Neuroeng Rehabil*, 7: 7. ISSN 1743-0003.

Amado, S., Simoes, M. J., Armada da Silva, P. A., Luis, A. L., Shirosaki, Y., Lopes, M. A., Santos, J. D., Fregnan, F., Gambarotta, G., Raimondo, S., Fornaro, M., Veloso, A. P., Varejao, A. S., Mauricio, A. C., & Geuna, S. (2008). Use of hybrid chitosan membranes and N1E-115 cells for promoting nerve regeneration in an axonotmesis rat model. *Biomaterials*, 29(33): 4409-4419. ISSN 0142-9612.

Amano, T., Richelson, E., & Nirenberg, M. (1972). Neurotransmitter synthesis by neuroblastoma clones (neuroblast differentiation-cell culture-choline acetyltransferase-acetylcholinesterase-tyrosine hydroxylase-axons-dendrites). *Proc Natl Acad Sci U S A*, 69(1): 258-263. ISSN 0027-8424.

Archibald, S. J., Shefner, J., Krarup, C., & Madison, R. D. (1995). Monkey median nerve repaired by nerve graft or collagen nerve guide tube. *J Neurosci*, 15(5 Pt 2): 4109-4123. ISSN 0270-6474.

Bain, J. R., Mackinnon, S. E., & Hunter, D. A. (1989). Functional evaluation of complete sciatic, peroneal, and posterior tibial nerve lesions in the rat. *Plast Reconstr Surg*, 83(1): 129-138. ISSN 0032-1052.

Baksh, D., Yao, R., & Tuan, R. S. (2007). Comparison of proliferative and multilineage differentiation potential of human mesenchymal stem cells derived from umbilical cord and bone marrow. *Stem Cells*, 25(6): 1384-1392. ISSN 1384-1392.

Beer, G. M., Steurer, J., & Meyer, V. E. (2001). Standardizing nerve crushes with a non-serrated clamp. *J Reconstr Microsurg*, 17(7): 531-534. ISSN 0743-684X.

Bertleff, M. J., Meek, M. F., & Nicolai, J. P. (2005). A prospective clinical evaluation of biodegradable neurolac nerve guides for sensory nerve repair in the hand. *J Hand Surg Am*, 30(3): 513-518. ISSN 0363-5023.

Bervar, M. (2000). Video analysis of standing--an alternative footprint analysis to assess functional loss following injury to the rat sciatic nerve. *J Neurosci Methods*, 102(2): 109-116. ISSN 0165-0270.

Bichara, D. A., Zhao, X., Hwang, N. S., Bodugoz-Senturk, H., Yaremchuk, M. J., Randolph, M. A., & Muratoglu, O. K. (2010). Porous poly(vinyl alcohol)-alginate gel hybrid construct for neocartilage formation using human nasoseptal cells. *J Surg Res*, 163(2): 331-336. ISSN 1095-8673.

Bini, T. B., Gao, S., Xu, X., Wang, S., Ramakrishna, S., & Leong, K. W. (2004). Peripheral nerve regeneration by microbraided poly(L-lactide-co-glycolide) biodegradable polymer fibers. *J Biomed Mater Res A*, 68(2): 286-295. ISSN 1549-3296.

Campbell, J. N. (2001). Nerve lesions and the generation of pain. *Muscle Nerve*, 24(10): 1261-1273. ISSN 0148-639X.

Can, A., & Karahuseyinoglu, S. (2007). Concise review: human umbilical cord stroma with regard to the source of fetus-derived stem cells. *Stem Cells*, 25(11): 2886-2895. ISSN 1549-4918.

Ceballos, D., Navarro, X., Dubey, N., Wendelschafer-Crabb, G., Kennedy, W. R., & Tranquillo, R. T. (1999). Magnetically aligned collagen gel filling a collagen nerve guide improves peripheral nerve regeneration. *Exp Neurol*, 158(2): 290-300. ISSN 0014-4887.

Chamberlain, L. J., Yannas, I. V., Hsu, H. P., & Spector, M. (2000). Connective tissue response to tubular implants for peripheral nerve regeneration: the role of myofibroblasts. *J Comp Neurol*, 417(4): 415-430. ISSN 0021-9967.

Chang, Y. H., Auyang, A. G., Scholz, J. P., & Nichols, T. R. (2009). Whole limb kinematics are preferentially conserved over individual joint kinematics after peripheral nerve injury. *J Exp Biol*, 212(Pt 21): 3511-3521. ISSN 3511-3521.

Chen, X., Wang, X. D., Chen, G., Lin, W. W., Yao, J., & Gu, X. S. (2006). Study of in vivo differentiation of rat bone marrow stromal cells into schwann cell-like cells. *Microsurgery*, 26(2): 111-115. ISSN 0738-1085.

Chen, Z. L., Yu, W. M., & Strickland, S. (2007). Peripheral regeneration. *Annu Rev Neurosci*, 30: 209-233. ISSN 0147-006X.

Conconi, M. T., Burra, P., Di Liddo, R., Calore, C., Turetta, M., Bellini, S., Bo, P., Nussdorfer, G. G., & Parnigotto, P. P. (2006). CD105(+) cells from Wharton's jelly show in vitro and in vivo myogenic differentiative potential. *Int J Mol Med*, 18(6): 1089-1096. ISSN 1107-3756.

Cui, L., Jiang, J., Wei, L., Zhou, X., Fraser, J. L., Snider, B. J., & Yu, S. P. (2008). Transplantation of embryonic stem cells improves nerve repair and functional recovery after severe sciatic nerve axotomy in rats. *Stem Cells*, 26(5): 1356-1365. ISSN 1549-4918.

de Medinaceli, L., Freed, W. J., & Wyatt, R. J. (1982). An index of the functional condition of rat sciatic nerve based on measurements made from walking tracks. *Exp Neurol*, 77(3): 634-643. ISSN 0014-4886.

Dellon, A. L., & Mackinnon, S. E. (1989). Sciatic nerve regeneration in the rat. Validity of walking track assessment in the presence of chronic contractures. *Microsurgery*, 10(3): 220-225. ISSN 0738-1085.

Den Dunnen, W. F., Meek, M. F., Robinson, P. H., & Schakernraad, J. M. (1998). Peripheral nerve regeneration through P(DLLA-epsilon-CL) nerve guides. *J Mater Sci Mater Med*, 9(12): 811-814. ISSN 0957-4530.

Fields, R. D., Le Beau, J. M., Longo, F. M., & Ellisman, M. H. (1989). Nerve regeneration through artificial tubular implants. *Prog Neurobiol*, 33(2): 87-134. ISSN 0301-0082.

Fu, Y. S., Shih, Y. T., Cheng, Y. C., & Min, M. Y. (2004). Transformation of human umbilical mesenchymal cells into neurons in vitro. *J Biomed Sci*, 11(5): 652-660. ISSN 1021-7770.

Geuna, S. (2005). The revolution of counting "tops": two decades of the disector principle in morphological research. *Microsc Res Tech*, 66(5): 270-274. ISSN 1059-910X.

Geuna, S., Raimondo, S., Ronchi, G., Di Scipio, F., Tos, P., Czaja, K., & Fornaro, M. (2009). Chapter 3: Histology of the peripheral nerve and changes occurring during nerve regeneration. *Int Rev Neurobiol*, 87: 27-46. ISSN 0074-7742.

Geuna, S., Tos, P., Guglielmone, R., Battiston, B., & Giacobini-Robecchi, M. G. (2001). Methodological issues in size estimation of myelinated nerve fibers in peripheral nerves. *Anat Embryol (Berl)*, 204(1): 1-10. ISSN 0340-2061.

Gramsbergen, A., Ijkema-Paassen, J., & Meek, M. F. (2000). Sciatic Nerve Transection in the Adult Rat: Abnormal EMG Patterns during Locomotion by Aberrant Innervation of Hindleg Muscles. *Experimental Neurology*, 161(1): 183-193. ISSN 0014-4886.

Grant, C., Twigg, P., Egan, A., Moody, A., Smith, A., Eagland, D., Crowther, N., & Britland, S. (2006). Poly(vinyl alcohol) hydrogel as a biocompatible viscoelastic mimetic for articular cartilage. *Biotechnol Prog*, 22(5): 1400-1406. ISSN 8756-7938.

Grimpe, B., & Silver, J. (2002). The extracellular matrix in axon regeneration. *Prog Brain Res*, 137: 333-349. ISSN 0079-6123.

Gu, X., Ding, F., Yang, Y., & Liu, J. (2011). Construction of tissue engineered nerve grafts and their application in peripheral nerve regeneration. *Prog Neurobiol*, 93(2): 204-230. ISSN 1873-5118.

Hood, B., Levene, H. B., & Levi, A. D. (2009). Transplantation of autologous Schwann cells for the repair of segmental peripheral nerve defects. *Neurosurg Focus*, 26(2): E4. ISSN 1092-0684.

Howard, C. S., Blakeney, D. C., Medige, J., Moy, O. J., & Peimer, C. A. (2000). Functional assessment in the rat by ground reaction forces. *J Biomech*, 33(6): 751-757. ISSN 0021-9290.

Hu, D., Hu, R., & Berde, C. (1997). Neurologic evaluation of infant and adult rats before and after sciatic nerve blockade. *Anesthesiology*, 86: 957-965. ISSN 003-3022.

Huang, Y. C., & Huang, Y. Y. (2006). Biomaterials and strategies for nerve regeneration. *Artif Organs*, 30(7): 514-522. ISSN 0160-564X.

Itoh, S., Suzuki, M., Yamaguchi, I., Takakuda, K., Kobayashi, H., Shinomiya, K., & Tanaka, J. (2003). Development of a nerve scaffold using a tendon chitosan tube. *Artif Organs*, 27(12): 1079-1088. ISSN 0160-564X.

Joao, F., Amado, S., Veloso, A., Armada-da-Silva, P., & Mauricio, A. C. (2010). Anatomical reference frame versus planar analysis: implications for the kinematics of the rat hindlimb during locomotion. *Rev Neurosci*, 21(6): 469-485. ISSN 0334-1763.

Johnson, E. O., & Soucacos, P. N. (2008). Nerve repair: experimental and clinical evaluation of biodegradable artificial nerve guides. *Injury*, 39 Suppl 3: S30-36. ISSN 1879-0267.

Johnson, W. L., Jindrich, D. L., Roy, R. R., & Reggie Edgerton, V. (2008). A three-dimensional model of the rat hindlimb: musculoskeletal geometry and muscle moment arms. *J Biomech*, 41(3): 610-619. ISSN 0021-9290.

Kadivar, M., Khatami, S., Mortazavi, Y., Shokrgozar, M. A., Taghikhani, M., & Soleimani, M. (2006). In vitro cardiomyogenic potential of human umbilical vein-derived mesenchymal stem cells. *Biochem Biophys Res Commun*, 340(2): 639-647. ISSN 0006-291X.

Karahuseyinoglu, S., Kocaefe, C., Balci, D., Erdemli, E., & Can, A. (2008). Functional structure of adipocytes differentiated from human umbilical cord stroma-derived stem cells. *Stem Cells*, 26(3): 682-691. ISSN 1549-4418.

Kerns, J. M., Braverman, B., Mathew, A., Lucchinetti, C., & Ivankovich, A. D. (1991). A comparison of cryoprobe and crush lesions in the rat sciatic nerve. *Pain*, 47(1): 31-39. ISSN 0304-3959.

Koka, R., & Hadlock, T. A. (2001). Quantification of functional recovery following rat sciatic nerve transection. *Exp Neurol*, 168(1): 192-195. ISSN 0014-4886.

Kranenburg, O., Scharnhorst, V., Van der Eb, A. J., & Zantema, A. (1995). Inhibition of cyclin-dependent kinase activity triggers neuronal differentiation of mouse neuroblastoma cells. *J Cell Biol*, 131(1): 227-234. ISSN 0021-9525.

Luis, A. L., Amado, S., Geuna, S., Rodrigues, J. M., Simoes, M. J., Santos, J. D., Fregnan, F., Raimondo, S., Veloso, A. P., Ferreira, A. J., Armada-da-Silva, P. A., Varejao, A. S., & Mauricio, A. C. (2007a). Long-term functional and morphological assessment of a standardized rat sciatic nerve crush injury with a non-serrated clamp. *J Neurosci Methods*, 163(1): 92-104. ISSN 0165-0270.

Luis, A. L., Rodrigues, J. M., Amado, S., Veloso, A. P., Armada-Da-Silva, P. A., Raimondo, S., Fregnan, F., Ferreira, A. J., Lopes, M. A., Santos, J. D., Geuna, S., Varejao, A. S., & Mauricio, A. C. (2007b). PLGA 90/10 and caprolactone biodegradable nerve guides for the reconstruction of the rat sciatic nerve. *Microsurgery*, 27(2): 125-137. ISSN 0738-1085.

Luis, A. L., Rodrigues, J. M., Geuna, S., Amado, S., Simoes, M. J., Fregnan, F., Ferreira, A. J., Veloso, A. P., Armada-da-Silva, P. A., Varejao, A. S., & Mauricio, A. C. (2008). Neural cell transplantation effects on sciatic nerve regeneration after a standardized crush injury in the rat. *Microsurgery*, 28(6): 458-470. ISSN 1098-2752.

Lundborg, G. (2002). Enhancing posttraumatic nerve regeneration. *J Peripher Nerv Syst*, 7(3): 139-140. ISSN 1085-9489.

Lundborg, G., Dahlin, L., Danielsen, N., & Zhao, Q. (1994). Trophism, tropism, and specificity in nerve regeneration. *J Reconstr Microsurg*, 10(5): 345-354. ISSN 0743-684X.

Mackinnon, S. E., Hudson, A. R., & Hunter, D. A. (1985). Histologic assessment of nerve regeneration in the rat. *Plast Reconstr Surg*, 75(3): 384-388. ISSN 0032-1052.

Marcus, A. J., & Woodbury, D. (2008). Fetal stem cells from extra-embryonic tissues: do not discard. *J Cell Mol Med*, 12(3): 730-742. ISSN 1582-1838.

Marreco, P. R., da Luz Moreira, P., Genari, S. C., & Moraes, A. M. (2004). Effects of different sterilization methods on the morphology, mechanical properties, and cytotoxicity of chitosan membranes used as wound dressings. *J Biomed Mater Res B Appl Biomater*, 71(2): 268-277. ISSN 1552-4973.

Masters, D., Berde, C., Dutta, S., Griggs, C., Hu, D., Kupsky, W., & Langer, R. (1993). Prolonged regional nerve blockade by controlled release of local anesthetic from a biodegradable polymer matrix. *Anesthesiology*, 79: 340-346.

Matsuyama, T., Mackay, M., & Midha, R. (2000). Peripheral nerve repair and grafting techniques: a review. *Neurol Med Chir (Tokyo)*, 40(4): 187-199. ISSN 0470-8105.

May, M. (1983). Trauma to the facial nerve. *Otolaryngol Clin North Am*, 16(3): 661-670. ISSN 0030-6665.

Meek, M. F., & Coert, J. H. (2002). Clinical use of nerve conduits in peripheral-nerve repair: review of the literature. *J Reconstr Microsurg*, 18(2): 97-109. ISSN 0743-684X.

Meek, M. F., Jansen, K., Steendam, R., van Oeveren, W., van Wachem, P. B., & van Luyn, M. J. (2004). In vitro degradation and biocompatibility of poly(DL-lactide-epsilon-caprolactone) nerve guides. *J Biomed Mater Res A*, 68(1): 43-51. ISSN 1549-3296.

Mligiliche, N. L., Tabata, Y., Kitada, M., Endoh, K., Okamato, K., Fujimoto, E., & Ide, C. (2003). Poly lactic acid--caprolactone copolymer tube with a denatured skeletal muscle segment inside as a guide for peripheral nerve regeneration: a morphological and electrophysiological evaluation of the regenerated nerves. *Anat Sci Int*, 78(3): 156-161. ISSN 1447-6959.

Papalia, I., Tos, P., Scevola, A., Raimondo, S., & Geuna, S. (2006). The ulnar test: a method for the quantitative functional assessment of posttraumatic ulnar nerve recovery in the rat. *J Neurosci Methods*, 154(1-2): 198-203. ISSN 0165-0270.

Pereira, J. E., Cabrita, A. M., Filipe, V. M., Bulas-Cruz, J., Couto, P. A., Melo-Pinto, P., Costa, L. M., Geuna, S., Mauricio, A. C., & Varejao, A. S. (2006). A comparison analysis of hindlimb kinematics during overground and treadmill locomotion in rats. *Behav Brain Res*, 172(2): 212-218. ISSN 0166-4328.

Pietak, A. M., Reid, J. W., Stott, M. J., & Sayer, M. (2007). Silicon substitution in the calcium phosphate bioceramics. *Biomaterials*, 28(28): 4023-4032. ISSN 0142-9612.

Raimondo, S., Fornaro, M., Di Scipio, F., Ronchi, G., Giacobini-Robecchi, M. G., & Geuna, S. (2009). Chapter 5: Methods and protocols in peripheral nerve regeneration experimental research: part II-morphological techniques. *Int Rev Neurobiol*, 87: 81-103. ISSN 0074-7742.

Rodrigues, J. M., Luis, A. L., Lobato, J. V., Pinto, M. V., Faustino, A., Hussain, N. S., Lopes, M. A., Veloso, A. P., Freitas, M., Geuna, S., Santos, J. D., & Mauricio, A. C. (2005). Intracellular Ca2+ concentration in the N1E-115 neuronal cell line and its use for peripheric nerve regeneration. *Acta Med Port*, 18(5): 323-328. ISSN 1646-0758.

Rodriguez JF, V.-C. A., Navarro X. (2004). Regeneration and funcional recovery following peripheral nerve injury. *Drugs Discovery Today: Disease Models*, 1(2): 177-185.

Sabatier, M. J., To, B. N., Nicolini, J., & English, A. W. (2011). Effect of slope and sciatic nerve injury on ankle muscle recruitment and hindlimb kinematics during walking in the rat. *J Exp Biol*, 214(Pt 6): 1007-1016. ISSN 1007-1016.

Schmidt, C. E., & Leach, J. B. (2003). Neural tissue engineering: strategies for repair and regeneration. *Annu Rev Biomed Eng*, 5: 293-347. ISSN 1523-9829.

Senel, S., & McClure, S. J. (2004). Potential applications of chitosan in veterinary medicine. *Advanced Drug Delivery Reviews*, 56(10): 1467-1480.

Shen, N., & Zhu, J. (1995). Application of sciatic functional index in nerve functional assessment. *Microsurgery*, 16(8): 552-555. ISSN 0738-1085.

Shirosaki, Y., Tsuru, K., Hayakawa, S., Osaka, A., Lopes, M. A., Santos, J. D., & Fernandes, M. H. (2005). In vitro cytocompatibility of MG63 cells on chitosan-organosiloxane hybrid membranes. *Biomaterials*, 26(5): 485-493.

Simoes, M. J., Amado, S., Gartner, A., Armada-Da-Silva, P. A., Raimondo, S., Vieira, M., Luis, A. L., Shirosaki, Y., Veloso, A. P., Santos, J. D., Varejao, A. S., Geuna, S., & Mauricio, A. C. (2010). Use of chitosan scaffolds for repairing rat sciatic nerve defects. *Ital J Anat Embryol*, 115(3): 190-210. ISSN 1122-6714.

Smith, K. J., & Hall, S. M. (1988). Peripheral demyelination and remyelination initiated by the calcium-selective ionophore ionomycin: in vivo observations. *J Neurol Sci*, 83(1): 37-53. ISSN 0022-510X.

Sporel-Ozakat, R. E., Edwards, P. M., Hepgul, K. T., Savas, A., & Gispen, W. H. (1991). A simple method for reducing autotomy in rats after peripheral nerve lesions. *Journal of Neuroscience Methods*, 36(2-3): 263-265.

Tateishi, T., Chen, G., & Ushida, T. (2002). Biodegradable porous scaffolds for tissue engineering. *J Artif Organs*, 5: 77-83.

Thalhammer, J. G., Vladimirova, M., Bershadsky, B., & Strichartz, G. R. (1995). Neurologic evaluation of the rat during sciatic nerve block with lidocaine. *Anesthesiology*, 82(4): 1013-1025. ISSN 0003-3022.

Trump, B. F., & Berezesky, I. K. (1995). Calcium-mediated cell injury and cell death. *FASEB J*, 9(2): 219-228. ISSN 0892-6638.

Tsien, R. Y. (1989). Fluorescent probes of cell signaling. *Annu Rev Neurosci*, 12: 227-253. ISSN 0147-006X.

Varejao, A. S., Cabrita, A. M., Geuna, S., Patricio, J. A., Azevedo, H. R., Ferreira, A. J., & Meek, M. F. (2003a). Functional assessment of sciatic nerve recovery: biodegradable poly (DLLA-epsilon-CL) nerve guide filled with fresh skeletal muscle. *Microsurgery*, 23(4): 346-353. ISSN 0738-1085.

Varejao, A. S., Cabrita, A. M., Meek, M. F., Bulas-Cruz, J., Filipe, V. M., Gabriel, R. C., Ferreira, A. J., Geuna, S., & Winter, D. A. (2003b). Ankle kinematics to evaluate functional recovery in crushed rat sciatic nerve. *Muscle Nerve*, 27(6): 706-714. ISSN 0148-639X.

Varejao, A. S., Cabrita, A. M., Meek, M. F., Bulas-Cruz, J., Gabriel, R. C., Filipe, V. M., Melo-Pinto, P., & Winter, D. A. (2002). Motion of the foot and ankle during the stance phase in rats. *Muscle Nerve*, 26(5): 630-635. ISSN 1652-1670.

Waitayawinyu, T., Parisi, D. M., Miller, B., Luria, S., Morton, H. J., Chin, S. H., & Trumble, T. E. (2007). A comparison of polyglycolic acid versus type 1 collagen bioabsorbable nerve conduits in a rat model: an alternative to autografting. *J Hand Surg Am*, 32(10): 1521-1529. ISSN 0363-5023.

Wang, H. B., Mullins, M. E., Cregg, J. M., Hurtado, A., Oudega, M., Trombley, M. T., & Gilbert, R. J. (2009). Creation of highly aligned electrospun poly-L-lactic acid fibers for nerve regeneration applications. *J Neural Eng*, 6(1): 016001. ISSN 1751-2560.

Wang, H. S., Hung, S. C., Peng, S. T., Huang, C. C., Wei, H. M., Guo, Y. J., Fu, Y. S., Lai, M. C., & Chen, C. C. (2004). Mesenchymal stem cells in the Wharton's jelly of the human umbilical cord. *Stem Cells*, 22(7): 1330-1337. ISSN 1330-1337.

Wang, W., Itoh, S., Matsuda, A., Ichinose, S., Shinomiya, K., Hata, Y., & Tanaka, J. (2008). Influences of mechanical properties and permeability on chitosan nano/microfiber mesh tubes as a scaffold for nerve regeneration. *J Biomed Mater Res A*, 84(2): 557-566. 1549-3296.

Weiss, M. L., Medicetty, S., Bledsoe, A. R., Rachakatla, R. S., Choi, M., Merchav, S., Luo, Y., Rao, M. S., Velagaleti, G., & Troyer, D. (2006). Human umbilical cord matrix stem cells: preliminary characterization and effect of transplantation in a rodent model of Parkinson's disease. *Stem Cells*, 24(3): 781-792. ISSN 1066-5099.

Wu, L. F., Wang, N. N., Liu, Y. S., & Wei, X. (2009). Differentiation of Wharton's jelly primitive stromal cells into insulin-producing cells in comparison with bone marrow mesenchymal stem cells. *Tissue Eng Part A*, 15(10): 2865-2873. ISSN 1937-335X.

Yang, C. C., Shih, Y. H., Ko, M. H., Hsu, S. Y., Cheng, H., & Fu, Y. S. (2008). Transplantation of human umbilical mesenchymal stem cells from Wharton's jelly after complete transection of the rat spinal cord. *PLoS One*, 3(10): e3336. ISSN 1932-66203.

Yoshii, S., & Oka, M. (2001). Peripheral nerve regeneration along collagen filaments. *Brain Res*, 888(1): 158-162. ISSN 0006-8993.

9

The Use of Biomaterials to Treat Abdominal Hernias

Luciano Zogbi
Federal University of Rio Grande- FURG
Brazil

1. Introduction

Hernia is the most frequent abdominal surgery. Although hernia is a highly prevalent disease, with serious risks and well-known anatomy, there were high rates of recurrence after treatment until the mid-20th century. With the advent of biomaterials, also called prostheses or meshes, the definitive cure of this disease is close to 100%. Using an appropriate surgical technique, following the appropriate postoperative care, with good integration of the prosthesis, a person can safely return to normal life with all the usual activities, including sports and physical efforts. Meshes are indicated for the treatment of all kinds of abdominal wall hernias, such as umbilical, epigastric, femoral and, mainly, inguinal and incisional hernias. Just as the surgical technique to treat this disease has evolved with a significant number of modalities, so also research and the prostheses market are taking up an increasingly outstanding position in the world (Usher, 1958; Penttinen & Grönroos, 2008).

2. Abdominal wall hernias

2.1 Definition

Hernia is derived from the Latin word for rupture. A hernia is defined as an abnormal protrusion of an organ or tissue through a defect, an opening, in its surrounding walls (figs. 1 and 2). This opening is called hernial ring. Its content may be any abdominal viscera, most frequently the small bowel and omentum. When protruding through the hernial ring, the herniated structure is covered by the parietal peritoneum, here called hernial sac (Malangoni & Rosen, 2007).

2.2 Classification

Although hernias can occur in various regions of the body, the most common site is the abdominal wall, particularly in the inguinal and ventral regions. **Hernias of the inguinal region** are classified as direct, indirect and femoral hernia, depending on where the hernia orifice is located in the fascia transversalis, in the deep inguinal ring and in the femoral ring, respectively. A **ventral hernia** is defined by a protrusion through the anterior abdominal wall fascia. These defects can be categorized as spontaneous or acquired or by their location on the abdominal wall: epigastric hernia occurs from the xyphoid process to the umbilicus; umbilical hernia occurs at the umbilicus; hypogastric hernia is a rare spontaneous hernia that occurs below the umbilicus in the midline; and acquired hernia typically occurs after

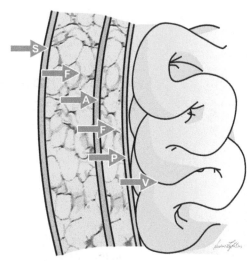

Fig. 1. Schematic drawing of a normal abdominal wall and their layers: Skin (S); Fat Tissue (F); Aponeurosis (A); Pre-peritoneal Fat Tissue (F); Peritoneun (P); and the abdominal viscera (V).

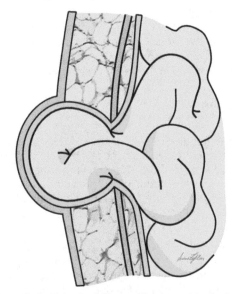

Fig. 2. Schematic drawing of a hernia. In this case, the bowel is the herniated viscera.

surgical incisions (figs 3 and 4). This is therefore termed *incisional hernia* and is the most common long-term complication after abdominal surgery (Franklin et al, 2003; Malangoni & Rosen, 2007; Penttinen & Grönroos, 2008).

Independently of the site, the principles of treatment are the same, only the surgical technique is different, according to regional anatomy.

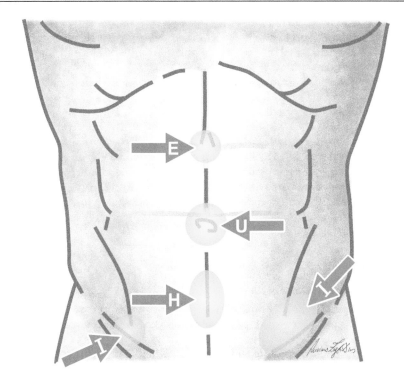

Fig. 3. Mainly places of abdominal wall hernia: Epigastric (E); Umbilical (U); Hypogastric (H); Inguinal (I).

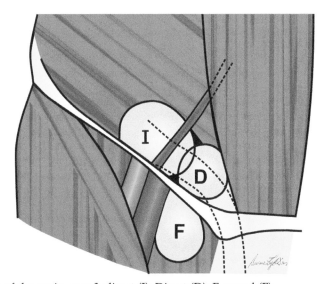

Fig. 4. Hernias of the groin area: Indirect (I); Direct (D); Femoral (F).

2.3 Epidemiology

Hernias are a common problem; however, their true incidence and prevalence are unknown. It is estimated that 5% of the population will develop an abdominal wall hernia, but the prevalence may be even higher. About 75% of all hernias occur in the inguinal region. Two thirds of these are indirect, and the remainder are direct inguinal hernias. The chance of a person having to undergo an inguinal hernia operation during his/her life is quite high, 27% in the case of men and 3% in the case of women. Men are 25 times more likely to have a groin hernia than are women. An indirect inguinal hernia is the most common hernia, regardless of gender. In men, indirect hernias predominate over direct hernias at a ratio of 2:1. Direct hernias are very uncommon in women. Although femoral hernias occur more frequently in women than in men, indirect inguinal hernias remain the most common hernia in women. About 3% to 5% of the population have epigastric hernias, and they are two to three times more common in men. The female-to-male ratio in femoral and umbilical hernias, however, is about 10:1 and 2:1, respectively (Malangoni & Rosen, 2007).

2.4 Risk factors

Hernias are characterized by the rupture of a wall that should be whole (incisional and inguinal direct hernias), or by the widening of a natural orifice (umbilical, femoral and direct inguinal hernias), generally due to excessive and sudden pressure on a fragile area. This weakening occurs because of biochemical and systemic changes in the collagen metabolism, weakening all the connective tissue. The best known risk factors are: old age, male sex, malnutrition, obesity, chemotherapy, radiotherapy, cortisone, sedentarism, decompensated diabetes mellitus, lack of vitamin C, anemia, smoking, chronic obstructive pulmonary disease, abdominal aortic aneurysm, long-term heavy lifting work, positive family history, pregnancy, appendicectomy, prostatectomy, peritoneal dialysis (Rodrigues et al., 2002; Wolwacz et al., 2003; Chan & Chan, 2005; Junge et al., 2006; Szczesny et al., 2006, Penttinen & Grönroos, 2008; Simons et al., 2009).

2.5 Complications if untreated

There is no spontaneous cure or medication to treat this disease. The only existing treatment is surgical correction. As long as it is not treated, the hernia defect will tend to become wider and increase progressively. Besides, herniated organs could be trapped by the hernial ring, and be unable to return to their usual site. When this happens it is called an **incarcerated** hernia. The risk of an inguinal hernia becoming incarcerated is less than 3% per year. The risk is greater in femoral hernias. The most serious risk of this disease is **strangulation**, which occurs when the incarcerated organ is deprived of a blood supply and becomes ischemic (fig.5). In this case, if the hernia is not treated urgently, its content may develop necrosis, infection, sepsis and death. When there is incarceration, the hernia must be reduced manually within 4 to 6 hours. After that, emergency surgery must be performed (Speranzini & Deutsch, 2001a).

Hernia surgery should ideally be performed electively, before these complications arise, making the procedure more effective and safe, since an emergency operation due to a strangulated inguinal hernia has a higher associated mortality (>5%) than an elective operation (<0.5%). Mortality increases about seven-fold after emergency operations and 20-fold if bowel resection was undertaken. After treatment, the risk of incarceration and/or strangulation disappears, as long as the hernia does not recur (Simons et al., 2009).

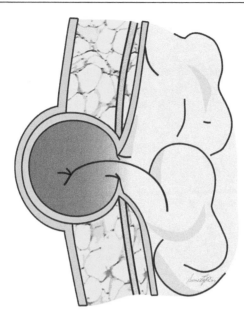

Fig. 5. Ischemic bowel due to strangulation

2.6 Treatment options

Although the only treatment is surgery, there are many effective surgical alternatives. However, merely correcting the hernia defect with sutures does not avoid the source of the problem, because the patient's tissues will still be fragile and predisposed to rupturing again at the same site. The recurrence rate for ventral hernia may be as high as 40–54% after open repair without meshes. Mesh repair is superior to suture repair, results in a lower recurrence rate and less abdominal pain. It does not cause more complications than suture repair (Burger et al. 2004; Penttinen & Grönroos, 2008).

For each type of hernia there are several techniques involving prostheses and different models of prosthesis. Surgeons in training, who see a variety of prosthetics in use, must recognize that the technique of prosthetic implantation is far more important than the type of prosthetic. To help the surgeon choose, it is helpful to look at the prosthetic landscape with a perspective based on (1) the prosthetic's raw material and design, (2) the implantation technique, and (3) the clinical scenario (Earle & Mark, 2008).

For treating inguinal hernia, the use of a polypropylene prosthesis is the best technique. Eighty-five percent of the operations, overall, are performed using an open approach and 15% are performed endoscopically. The surgeon should discuss the advantages and disadvantages of each technique with the patient. Endoscopic inguinal hernia techniques result in a lower incidence of wound infection, hematoma formation and an earlier return to normal activities or work than the Lichtenstein technique. When only considering chronic pain, endoscopic surgery is superior to open mesh. However, endoscopic inguinal hernia techniques need general anesthesia, result in a longer operation time and a higher incidence of seroma than the Lichtenstein technique (Simons et al., 2009).

Independently of the technique employed, after covering the hernia site adequately, the mesh must be fixed to the abdominal wall in order to prevent it from folding over or

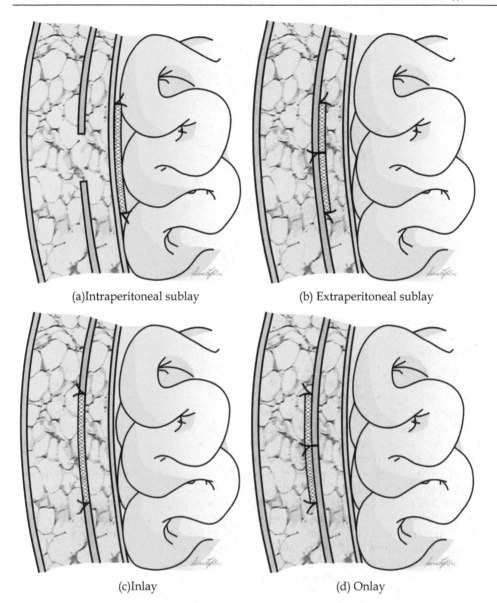

(a)Intraperitoneal sublay

(b) Extraperitoneal sublay

(c)Inlay

(d) Onlay

Fig. 6. Possible plans of the abdominal wall to insert the prosthesis.

migrating. It may be fixed simply by physical principles of pressure between layers of the abdominal wall (Stoppa & Rives, 1984), by means of a suture with inadsorbable thread (Lichtenstein et al., 1989), absorbable thread (Gianlupi & Trindade, 2004), clips (Read, 2011) or fibrin glue (Agresta & Bedin, 2008; Negro et al., 2011). For fixation of the mesh in ventral hernia repair, most authors have used an extraperitoneal – but intraperitoneal is also possible (fig.6a) - sub-lay technique, in which the mesh is sutured into place on the posterior rectus sheath with approximately 4 cm of fascia overlap (fig.6b). The other two repair options include an inlay technique (fig.6c), such that the mesh is sutured to the fascial edges, and an onlay technique whereby the mesh is placed and sutured onto the anterior rectus sheath (Fig.6d). The inlay technique has the advantage of minimal soft-tissue dissection thus reducing devascularized tissue, but the disadvantage of high rate of recurrences, while the onlay technique has the disadvantage of vast soft tissue dissection above the rectus layer (Penttinen & Grönroos, 2008).

3. History of biomaterials

Trusses have been used for the treatment of inguinal hernia for thousands of years. In the 19th century, several surgical techniques were proposed to treat hernia, but all of them limited themselves to the raphe of the hernia defect. Until then, the rate of occurrence was high, even surpassing 50%. Cooper already suspected of the degenerative nature of the disease and, Billroth, ahead of his time, perceived that even if he knew everything about anatomy and surgery, he still lacked something. Even before the meshes were created he said that: "If we could artificially produce tissues of the density and toughness of fascia and tendon, the secret for the radical cure of hernia would be discovered" (Amid, 1997; Franklin et al., 2003; Read, 2004; Earle & Mark, 2008).

The first biomaterials were described in 1900, when Oscar Witzel used a silver mesh (Witzel, 1900). Handley developed silk meshes in 1918, but they were no longer used because they did not tolerate the organism (Handley, 1963). In 1928, Goepel inserted stainless steel prostheses, of a fine, flexible, easily manipulated material (Goepel, 1900). Its drawback was the tendency to become fragmented, injuring tissue and blood vessels. The attempt to make celluloid-based materials, by Mandl, in 1933, did not meet with success, since, despite its flexibility and resistance to tension, it easily developed abscesses from infection (Mandl, 1962). In 1946, another metal material was described, vitalium, which was no longer used because of its rigidity (McNealy & Glassman, 1946). Amos Koontz adopted tantalum to treat eventrations in 1948, and it was widely accepted (Koontz, 1948). This was a resistant metal, with a low tendency to corrosion, appropriate to the synthesis of granulation tissue and very safe against infection. Its disadvantages were fragility and high cost, and therefore it was no longer used. The fragmentation observed in these metal substances over time is due to a principle of physics called *point of metal fatigue* (Sans,1986).

The era of plastics began in the manufacture of prostheses when nylon mesh was introduced in 1944 (Acquaviva & Bourret, 1944). Mersilene mesh, a polyester polymer, was widely known as an alloplastic material in 1946 (Adler & Firme, 1959). In 1951, Kneise described the use of the Perlon meshes (Kneise, 1953). In 1958, Francis Usher introduced the first generation of polyethylene mesh to correct abdominal hernias. Despite its good resistance and inertia, the clinical application of this material was limited because it could not be easily sterilized. In 1962 the same author fulfilled Billroth's dream and presented to the worldwide surgical community the material that, with Lichtenstein's encouraging results decades later, became the best known and most used: It is Marlex, a high density propylene; it cannot be affected by acids, alkalis, or organic solvents; it is highly resistant; inert to the infectious process; non-toxic; it cannot be absorbed; it can be cut and modeled without deforming; in other words, all of the benefits of polyethylene added to the virtue of being possible to sterilize in the autoclave (Usher, 1958, 1962; Lichtenstein, 1989). Mesh screens of other materials, also published at the time, such as chromed catgut (Schönbauer & Fanta, 1958),

Silastic, based on silicone (Brown et al., 1960), and Supramid (Rappert , 1963), were not successful (table 1).

Year	Author	Material
1900	Witzel	Silver
1918	Handley	Silk
1928	Goepel	Stainless steel
1933	Mandl	Celluloid
1944	Acquaviva & Bourret	Nylon
1946	Mc Nealy & Glassman	Vitalium
1946	Adler & Firme	Mersilene
1948	Koontz	Tantalum
1951	Kneise	Perlon
1958	Schönbauer & Fanta	Chromed catgut
1958	Usher	Polyethylene
1960	Brown et al.	Silastic
1962	Rappert	Supramid
1962	Usher	Marlex

Table 1. Development of synthetic prosthesis over the course of history.

4. Mechanism of biomaterial integration to the organism

4.1 Normal healing

After tissue injury, such as surgery, the healing process occurs. It takes place in three phases. It begins with the **inflammatory, substrate or exudative phase,** characterized firstly by vasoconstriction and platelet aggregation. Fibrin is formed as the coagulation mechanism continues, in order to diminish loss to hemorrhage, and it lasts approximately 15 minutes. Then the opposite phenomenon is observed, with the consequent exudation of proteins and plasma cells in the zone affected. The cell response is processed 6 to 16 h after the onset of the lesion, when a large amount of polymorphonuclear neutrophils appear, as the *first wave of cell migration*. They stay from 3 to 5 days, with a peak within 68 h (Monaco & Lawrence, 2003). Already on the 1st day there is a monocyte incursion. These are macrophage precursors. Neocapillary growth and fibroblastic proliferation begin about 36 h after injury. The activated macrophages are the predominant leukocytes on day 3, when they peak and persist until healing is complete. This first phase lasts until the 2nd day (Castro & Rodrigues, 2007), and may last until the 4th day postoperatively (Pitrez, 2003).

Around the 3rd to 5th day the **proliferative or connective tissue phase** begins, in which angiogenesis and fibroplasia occur, from the proliferation of the endothelial cells and fibroblasts, respectively. They will build the *granulation tissue*. The lymphocytes appear around the 5th day, peaking on the 7th day , and they are mostly represented by T Lymphocytes. During the 2nd week, the fibroblasts become the dominant cells, especially on the 10th day. After this period they differentiate into fibrocytes. Fibroblasts synthetize collagen, which promotes repair resistance. Around the second week type III collagen is gradually replaced by type I collagen. The fibroblasts migrate into the wound from the surrounding tissue, differentiating into myofibroblasts, forming actin filaments, synthetizing a collagen that is periodically reabsorbed, and like the muscles, the scar tissue has a centripetal movement, making the scar spheroid (Nien et al., 2003). Wound contraction is an essential aspect of healing. It diminishes the area of the defect making it easier to close. During this phase, tension resistance of the synthetized tissue is still low, no more than 25% to 30% of the original resistance (Junge et al., 2002; Klinge et al., 2002).

From the 21st day onwards, during the last phase of the healing process, called **molding, maturing, resolutive or differentiation phase,** tension resistance will reach its highest levels. The accumulation of collagen tissue peaks on the 21st day, and its value remains practically constant in the 3 following months. During this period, acute and inflammatory cells diminish, angiogenesis is suppressed, and fibroplasia ends. The balance between synthesis and degradation of collagen is restored, and *reformulation of collagens* is seen. In the mature matrix type I is 80% to 90%, and type III is 10 to 20% of the total collagen. This matrix undergoes continuous modification until a stable matrix is formed. The scar tissue takes on 40% of the tensile resistance around 6 weeks, 80% around 6 months, and its maximum resistance is achieved after many months, or even years, but it is not equal to the resistance of healthy tissue. (Monaco & Lawrence, 2003; Pitrez, 2003).

4.2 Healing with a prosthesis

The reinforcement given by the prosthesis does not occur due to the mere mechanical presence of the material at the surgical site. It is caused by the tissue that will be produced because it is there. After any prosthetic is implanted, an extraordinarily complex series of events takes place and the healing process described above will occur amidst the mesh. The architecture formed by its filaments and by its pores will act as a foundation for the deposition of connective tissue. The principle and phases of healing are similar, and on the mesh screen weave, a new tissue will be built similar to a dense aponeurosis (Zogbi et al., 2010).

Immediately after implantation, the prosthetic adsorbs proteins that create a coagulum around it. This coagulum consists of albumin, fibrinogen, plasminogen, complement factors, and immunoglobulins. Platelets adhere to this protein coagulum and release a host of chemoattractants that invite other platelets, polymorphonucleocytes (PMNs), fibroblasts, smooth muscle cells, and macrophages to the area in a variety of sequences. Activated PMNs drawn to the area release proteases to attempt to destroy the foreign body in addition to organisms and surrounding tissue. The presence of a prosthetic within a wound allows the sequestration of necrotic debris, slime-producing bacteria, and a generalized prolongation of the inflammatory response of platelets and PMNs. Macrophages then increasingly populate the area to consume foreign bodies as well as dead organisms and tissue. These cells ultimately coalesce into foreign body giant cells that stay in the area for an indefinite period of time (Earle & Mark, 2008). The histological examination of the mesh screens removed shows that all prostheses, independent of type of biomaterial, induce an acute and intense inflammatory reaction (Zogbi et al., 2010, whose quantity and quality depend on the type of material of which the mesh is made (Di Vita et al., 2000). The fibroblasts and smooth muscle cells subsequently secrete monomeric fibers that polymerize into the helical structure of collagen deposited in the extracellular space. The overall strength of this new collagen gradually increases for about 6 months, resulting in a relatively less elastic tissue that has only 70% to 80% of the strength of the native connective tissue. It is for this reason that the permanent strength of a prosthetic is important for the best long-term success of hernia repair (Earle & Mark, 2008).

Three aspects are valuable from the histological standpoint, in the interaction between the material and the organism: the size of the tissue reaction, the cell density and fibroblastic activity. The tissue reaction is 10mm on the 20th day and 20 mm on the 40th day. Cell censity is moderate to the 8th day and maximal after the 30th day. Fibroblastic activity begins on the 8th day on the intraperitoneal plane and 10th day on the extraperitoneal plane. It is maximal on days 30 and 35, respectively. The mechanical resistance of wall reconstruction is similar at the end of 30 days, independently of the material used. During the early postoperative period, between the first and second week, the permeable macroporous prostheses are significantly more resistant than the impermeable ones. This period, during which the prosthesis insertion zone is fragile, is called the Howes latency period (Sans, 1986; Zogbi et al., 2010).

5. Classification

Currently there are more than 70 meshes for hernia repair available on the market (Eriksen et al., 2007). They can be classified into different categories according to composition or type of material (Ponka, 1980), pore size (table 2) (Amid, 1997), density (Earle & Mark, 2008) and others. The classification below covers all these characteristics:

5.1 Synthetic nonabsorbable prosthesis
5.1.1 Type I: Totally macroporous prosthesis
The macroporous prostheses are characterized by a diameter larger than 75 (Amid, 1997) or 100µm (Annibballi & Fitzgibbons, 1994). Thus, they allow easy entry of macrophages, fibroblasts, collagen fibers, which will constitute the new connective tissue and integrate the prosthesis to the organism. They also allow more immunocompetent cells to enter, providing protection from infection-causing germs. The larger the pore diameter, the greater and faster will be the fibroplasia and angiogenesis (Gonzalez et al., 2005). On the other hand, there will also be a greater risk of adhesions when the screen is inserted in the intraperitoneal space, especially if it is in contact with the viscerae; it may also promote erosion and fistula formation (Hutchinson et al., 2004; Mathews et al, 2003; Melo et al., 2003). The main representative is **Polypropylene (PP)** (fig.7). Common brand names include **Marlex®** (Davol, Cranston, Rhode Island), **Prolene®** (Ethicon, Somerville, New Jersey), **Prolite®** (Atrium Medical, Hudson, New Hampshire), **Atrium®** and **Trelex®** (Erle & Mark, 2008). PP is the material most used to correct hernias, both anteriorly, retroperitoneally or laparoscopically (Bellón, 2009). PP is an ethylene with an attached methyl group, and it was developed and polymerized in 1954 by the Italian scientist, Giolo Natta. It is derived from propane gas. The position of the methyl groups during polymerization affects overall strength and it is at a maximum when they are all on the same side of the polymeric chain. This polymer is hydrophobic, electrostatically neutral, and resistant to significant biologic degradation (Earle & Mark, 2008). Since it is thermostable, with a fusion point of 335°F, it can be sterilized repeatedly in an autoclave (Amid, 2001). Studies show that the tensile strength of PP implanted in organic tissue remains unchanged over time. Disposed in different makes and models, the mesh screens developed for use in hernioplasties are monofilamentary, rough, semi-rigid and allow elasticity in both directions (Speranzini & Deutsch, 2001b). The screen thickness varies according to the model. For instance, Atrium®, Marlex®, Prolene® are respectively, 0.048, 0.066 and 0.065 cm (Goldstein, 1999). It has a high tolerance to infection. When there is a infection at the surgical site, the mesh screen can be

preserved, as long as it is integrated to the fascia, thanks to its broad pores, and must only be drained and the infection treated. In open inguinal hernia repair, the use of a monofilament polypropylene mesh is advised to reduce the chance of incurable chronic sinus formation or fistula which can occur in patients with a deep infection (Simons et al., 2009).

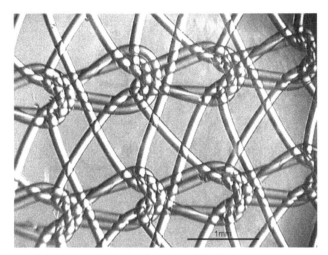

Fig. 7. Augmentation picture of polypropylene prosthesis

Considering the abdominal cavity as a cylinder, and according to Pascal's hydrostatic principle, the maximum load for its rupture is between 11 and 27N/cm. Abdominal pressures vary from 8 to 150mmHg. Klinge et al demostrated that the prostheses that were being used until that time can bear up to 10 times these rupture tensions, much higher than the resistance of the abdominal wall itself. Thus, there is a reduction of the natural elasticity in the aponeurosis after it is implanted, since the incorporation of tissue to the prosthesis gives rise to an incongruence of resistance between the receiving tissue and the biomaterial, and can cause patient more discomfort. Therefore, it would be more reasonable to implant materials with a lower resistance and greater elasticity (Bellón, 2009). Low weight density (LW) prostheses were then developed (fig.8), characterized by a lower concentration of synthetic material and larger pores (>1,000 µm). The first experimental tests were performed with a hybrid prosthesis of LW PP and polyglactine (Klinge et al., 1998), which was later sold under the name **Vypro II**® (Ethicon, Johnson&Johnson, Somerville, USA). Then pure LW PP prostheses were developed and disseminated, such as **Parietene**® (Tyco, Healthcare, Mansfield, MA), with a 38g/m² density and 1.15 +- 0.05 mm² pores and **Optilene elastic**® (Braun, Spangerwerg, Germany), with 48g/m² and 7.64 +- 0.32mm² pores (Bellón, 2009).

Hence, as to density, the prostheses can be classified as: Heavyweight (HW), when they are above 80g/m²; Mediumweight (MW), between 50 and 80 g/m²; Lightweight (LW), between 35 and 50 g/m²; and Ultra-lightweight, below 35 g/m². Comparing them, it would be helpful to classify density (weight) and pore size uniformly in a standard fashion. Earle & Mark proposed a standard based on currently available data: Very large pore: >2,000 µm; Large pore: 1,000–2,000 µm; Medium pore: 600–1,000 µm; Small pore: 100–600 µm; Microporous (solid): <100 µm (Earle & Mark, 2008).

Fig. 8. Comparison between a HW (Marlex®), on the left, and a LW mesh (Parietene®), on the right.

Material-reduced (Weight-reduced mesh materials/ lightweight/oligofilament structures/largepore/macroporous>1,000 μm) meshes have some advantages with respect to long-term discomfort and foreign-body sensation in open hernia repair (when only chronic pain is considered), but are possibly associated with an increased risk for hernia recurrence in high-risk conditions (large direct hernia), perhaps due to inadequate fixation and/or overlap. They seem to shrink less, cause less inflammatory reaction and induce less extensive scar-tissue formation (Hollinsky et al., 2008; Simons et al., 2009).

5.1.2 Type II: Totally microporous prosthesis

The pores are smaller than 10μm in at least one of the three sizes. The main example is **expanded politetraflouroethylene (e-PTFE)**. It was discovered at a DuPont laboratory serendipitously by Roy Plunkett in 1938. While researching tetraflouroethylene gas as a refrigerant, he discovered that the gas spontaneously polymerized into a slippery, white, powdery wax. After some time on the shelf, it was eventually used as a coating for cables. While still working at DuPont, William Gore subsequently saw the potential for medical applications, and ultimately started his own company, W.L. Gore and Associates, in 1958. That company developed and manufactured e-PTFE under the brand name **Gore-Tex®** (W.L. Gore and Associates, Flagstaff, Arizona) for hernia repair products, among other things. There are other manufacturers of PTFE hernia prosthetics, each with a different manufacturing process, and hence a slightly different architecture (Earle & Mark, 2008). PTFE is not a mesh, but a flexible, impervious sheet. It is transformed into its expanded form (e-PTFE) after being submitted to an industrial process. It is a soft, flexible, slightly elastic material, and its smooth surface is not very adherent (Mathews et al., 2003). Therefore it must be carefully fixed with sutures, since its integration is very slow, taking about 30 to 40 days (Speranzini & Deutsch, 2001b). Its minuscule pores are actually complex fine canals, through which fibroblasts penetrate and synthetize collagen. The e-PTFE is composed of columns of compact nodules, interconnected by fine fibers of the same material (Mathews et al., 2003). The intermodal distance is from 17 to 41μm, with a multidirectional fibrillar arrangement that provides equal strength on every plane (Amid, 2001). Bacteria, approximately 1μm in size, easily penetrate the micropores of the prosthesis and are thus protected from the macrophages or neutrophils, which are too voluminous to enter the site, perpetuating the infectious process. It is a mechanism similar to that of a foreign body which occurs with plaited threads or any materials with interstices (Amid, 1997). Therefore, when there is an infection, the mesh screen should always be removed, on the contrary of the macroporous screens. The main advantage of this material is the diminished risk of adhesions, even in direct contact with the viscerae. It is the prosthesis with the smallest tissue reaction (Speranzini & Deutsch, 2001b). Because of this, its use in laparoscopic hernia repair allows the surgeon to leave the peritoneum open once the prosthetic is in place (Earle & Mark, 2008).

5.1.3 Type III: Macroporous prosthesis with multifilament or microporous components
They are characterized as containing plaited multifilamentary threads in their composition, and the space between threads is less than 10 μm; but also because their pores are larger than 75 μm. They include plaited polyester mesh - **Mersilene®** (Ethicon, Johnson&Johnson, Somerville, USA) and **Parietex®** (Covidien, Mansfield, USA); plaited polypropylene - **SurgiPro®** (Covidien, Mansfield, USA); perforated PTFE - **Mycromesh®** and **MotifMESH®** (Amid, 1997; Eriksen et al., 2007).
The main disadvantage is during an infection, because the chance of complete wound healing after adequate drainage is difficult. When a multifilament mesh is used, bacteria (<1 μm) can hide from the leucocytes (>10 μm), because the mesh has a closer weave structure with a smaller pore diameter (<10 μm) (Simons et al., 2009).
Polyester (PE), the common textile term for polyethylene terephthalate (PET), is a combination of ethylene glycol and terephthalic acid, and it was patented by the English chemists J.R. Whinfield and J.T. Dickson in 1941 at the Calico Printers Association Ltd. in Lancashire, the United Kingdom. PET is hydrophilic and thus has the propensity to swell. PET is the same polymer used for plastic beverage bottles (Earle & Mark, 2008). It is a light, soft, flexible, elastic material, in a single direction. Its wide meshes encourage fibroblastic migration making it easier for tissue to incorporate – its pores are even greater than those of the PP, which is believed to allow faster cell migration and greater intensity of adherence to the underlying fascia (Gonzalez et al., 2005). It has good resistance to infection, although its threads are multifilament. It does not have the plastic memory of PP, which allows it to adapt to the structures on which it is placed. Another advantage is the cost, because it has a lower cost (Speranzini & Deutsch, 2001b). It is the mesh screen most used by European surgeons, especially the French (Stoppa & Rives, 1984).
In 1993 the MycroMesh® with pores all way through the mesh was introduced to allow better tissue ingrowth. MotifMESH® is a new macroporous non-woven mesh of condensed PTFE (cPTFE) for intraperitoneal application. Although the mesh is macroporous (fenestrated) it has a theoretically anti-adhesion barrier because of the PTFE content. The thickness of the MotifMESH® is reduced by 90% compared with older ePTFE meshes (Eriksen et al., 2007).

Prostheses	Definition	Examples
Type I	Pore diameter > 75 μm	Polypropyleno (PP)
Type II	Pore diameter < 10 μm	Expanded Polytetrafluorethylene (e-PTFE)
Type III	Pore diameter >75 μm Space between threads < 10 μm	Polyester (PE)
Type V	Submicromic pores	Pericardium, dura mater

Table 2. Classification of the biomaterials according to Amid (Amid, 1997)

5.2 Mixed prostheses
Also known as "second generation" screens, they are characterized by combining more than one type of material in the same prosthesis (Bachman & Ramchaw, 2008).

5.2.1 Partially absorbable prosthesis
One of the disadvantages of LW prostheses is the excessive malleability of the screen). The lack of memory, or lack of rigidity, makes them difficult to handle during surgery, especially laparoscopic surgery. To reduce the polymer density (and subsequent inflammatory response), yet maintain the intraoperative handling characteristics and long-term wound strength, prosthetics have been developed that mix nonabsorbable polymers (eg, PP) with absorbable polymers. Thus, screens composed by a LW PP structure are associated with biodegradable elements, such as **polyglactine** – **Vypro II®** mesh or **polyglecaprone-25 - Ultrapro®** mesh (Ethicon, Johnson&Johnson, Somerville, USA). This confers on the screen an appropriate malleability for better surgical handling, without, however, leaving a high weight of unabsorbed tissue in the organism (Earle & Romanelli, 2007; Earle & Mark, 2008; Hollinsky et al., 2008; Bellón, 2009).

5.2.2 Coated nonabsorbable prosthesis
In order to avoid visceral adhesions, erosion and even fistula formation which are possible complications of macroporous screens when inserted on the peritoneal side, screens covered with low tissue reaction material were developed to remain in direct contact with the viscerae. The two-sided **DualMesh®** was introduced in 1994, made in e-PTFE, and it was later modified with large interstices and an irregular "corduroy-like" surface on the parietal side to increase tissue ingrowth. Other available brands are: **Intramesh T1®**; **Dulex®**; and **Composix®**. The DualMesh® is also available with incorporated antimicrobial agents (silver–chlorhexidine film, type "Plus"). **TiMesh®** (GfE Medizintechnik GmbH, Nürnberg, Germany) is a titanium-coated lightweight (macroporous) PP mesh. Titanium is known for its good biocompatibility and should theoretically reduce adhesions. It is manufactured for intraperitoneal use although it has no "real" solid anti-adhesion barrier or micro-pore/no-pore site against the bowel loops. **Parietene Composite®** (Covidien, Mansfield, USA) is a woven PP mesh with a protective collagen-oxidized film (collagen-coating) on the visceral side. **Sepramesh®** is a PP mesh coated on the visceral side with an absorbable barrier of sodium hyaluronate and carboxymethylcellulose. **Proceed®** (Ethicon, Johnson&Johnson, Somerville, USA) is a Prolene® soft mesh encapsulated in a polydioxanone polymer film (PDS®) covered by a layer of absorbable oxidised regenerated cellulose (ORC); **Glucamesh®** (Brennen Medical, St. Paul, Minnesota) is a midweight PP mesh (50 g/m²) coated with the absorbable complex carbohydrate, oat beta glucan; **Dynamesh®** (FEG Textiltechnik, Aachen, Germany) is a PP mesh with polyvinylidene fluoride (PVDF) monofilament; **C-QUR®**

(Atrium Medical) is a mediumweight PP mesh (50 or 85 g/m^2) coated with an absorbable omega-3 fatty acid preparation derived from fish oil, because omega-3 fatty acids have anti-inflammatory properties. The coating is about 70% absorbed in 120 days and has had all protein removed to avoid an immune response. The same mesh without the coating has been analyzed in the laboratory and found to be acceptable in terms of inflammatory response compared with more heavyweight polypropylene prosthetics (Mathews et al., 2003; Abaza et al., 2005; Eriksen et al., 2007; Earle & Mark, 2008; Schreinemacher et al., 2009). The **Parietex Composite**® mesh is composed of multifilament PE with a resorbable collagen-oxidized film made of oxidized atelocollagen type I, polyethylene glycol and glycerol, against the viscera. **Intramesh W3**® is a PE mesh with silicone layer (Eriksen et al., 2007; Schreinemacher et al., 2009).

In the mixed prostheses in general, weight is usually smaller and porosity greater. For instance, a conventional PP prosthesis PP (HW) such as Surgipro®, weighs $84g/m^2$ and has small pores ($0.26 +- 0.03mm^2$). Conversely, Ultrapro®, weighs $28g/m^2$ with $3.45 +- 0.19mm^2$; pores; and VyproII® weighs $35g/m^2$ with $4.04 +- 0,54mm^2$ pores(Bellón, 2009).

5.3 Biologic prosthesis

Biologic mesh materials are based on collagen scaffolds derived from a donor source and they represent so-called "third-generation" mesh. According to Amid's classification they are included in the **type IV** prostheses, **biomaterials with submicronic pore size.** Dermis from human, porcine, and fetal bovine sources are decellularized to leave only the highly organized collagen architecture with the surrounding extracellular ground tissue. Other natural collagen sources in addition to the dermal products include porcine small intestine submucosa (which is layered for strength) and bovine pericardium. The collagen in these materials can be left in its natural state or chemically crosslinked to be more resistant to the collagenase produced in wounds. By increasing crosslinking, the persistence of the mesh is also increased. Uncrosslinked mesh can be totally incorporated and reabsorbed within 3 months, whereas a highly crosslinked mesh can persist for years (Amid, 1997; Bachman & Ramshaw, 2008).

Most of the human studies published on biologic materials are from difficult clinical situations. Because angiogenesis is a part of the remodeling of the mesh, these materials can potentially resist infection (Blatnik et al., 2008; Deprest et al., 2009), and they have a moderately good success rate for salvaging contaminated and infected fields, especially when placed with wide overlap. Other findings demonstrate some resistance to adhesion formation (Bachman & Ramchaw, 2008).

The basic concept behind these types of prosthetics is that they provide a matrix for native cells to populate and generate connective tissue that will replace the tissue in the hernia defect. Given that newly formed connective tissue is only 70% to 80% as strong as native connective tissue, and that hernia patients may have an inherent defect in their native connective tissue, biologic (or absorbable synthetic) prosthetics would theoretically have a higher risk of recurrence than would permanent prosthetics. With a theoretically increased risk of long-term recurrence, relatively high cost, and no clear benefit, the use of these products for elective inguinal hernia repair should be considered investigational, and are not routinely indicated (Earle & Mark, 2008; Simons et al., 2009).

5.3.1 Heterografts

These are biomaterials produced from animal tissue. Porcine heterografts, whose main and most studied example is **Surgisis**® (Cook Biomedical, Bloomington, IN, USA), derived from porcine small bowel submucosa was one of the initial biologic grafts used and was FDA approved in 1999. Surgisis is an acellular xenograft consisting primarily of type I porcine collagen. It is harvested from slaughterhouse pigs, processed with paracetic acid, and terminally sterilized with ethylene oxide. It does not undergo crosslinking during processing. The graft is available in different thickness including four ply, for hiatal hernias and groin hernias; and eight ply, for ventral hernias. This material seems to be biodegradable and manufacturers claim it is completely replaced with native tissue at 6 months. Surgisis has been extensively studied in animal models. Agresta & Bedin say that, besides diminishing the chances of adhesions on the peritoneal side, another advantage to using a biological mesh is that the persistence of a synthetic mesh in the preperitoneal inguinal area, where scar formation can result in the possible complications of infertility and difficulties in future vascular and urological surgical procedures, is that the biological mesh does not lead to a persistent foreign body in this region and these complications may, therefore, be avoided. This is especially important in the young patient or in athletes. The theoretical benefits of its use include: resistance to infection in contaminated surgical fields; avoidance of a permanent foreign body in the inguinal region; and a reconstruction that results in the formation of natural tissue. These characteristics, together with the results of several human clinical studies which have demonstrated its safety, let us suggest its possible use in young patients without any fear of possible future complications (Agresta & Bedin, 2008).

Several porcine dermal products are also available using different processing techniques: **Permacol®** (Covidien, Norwalk, CT) was initially approved for pelvic floor reconstruction in 2000. It is manufactured by Tissue Sciences Laboratory (Aldershot, UK) and was acquired by Covidien in 2008. It is processed with Diisocynate to achieve collagen cross-linking, and is terminally sterilized with gamma irradiation. It is not freeze dried and requires no rehydration before use. There are several peer reviewed publication evaluating Permacol in gynecologic, urological, plastic surgical, and ventral hernia repairs. **Collamend®**, distributed by Davol Inc, (Warwick, RI), was approved for use in 2006. It is freeze dried, requires 3 minutes of rehydration before use, and is heavily crosslinked. LifeCell Inc, (Branchburg, NJ) introduced **Strattice®** in 2007; it is, noncrosslinked, terminally sterilized, and requires a 2 minute soak before usage. **XenMatrix®** is another porcine dermal product that received approval in 2003, and is manufactured by Brennen Medical (St, Paul, MN). It is noncrosslinked and terminally sterilized with E-beam radiation. It is stored at room temperature and does not require rehydration before usage (Rosen, 2010).

Bovine donors constitute the remainder of the heterografts and sources include pericardium or fetal dermis. **Tutopatch®** is a bovine pericardial product, is manufactured by Tutogen (Alachua, FL), and received FDA approval in 2000. Tutogen processing is a proprietary technique that involves osmotic contrast bathing, hydrogen peroxide, sodium oxide, and gamma irradiation for terminal sterilization. It is stored at room temperature and requires rehydration before use. Two other bovine pericardial products are manufactured by Synovis Surgical Innovations (St, Paul, MN): **Veritas®** received approval in 2003, and is a noncrosslinked bovine pericardium that is harvested from an isolated Midwestern slaughterhouse from cows younger than 30 months. It is processed with sodium hydroxide, propylene oxide, and ethanol. It does not require rehydration and is ready to use out of the package. **Peri-guard®** is treated with gluteraldehyde to initiate collagen crosslinking. It requires a 2 min soak before usage (Rosen, 2010).

5.3.2 Allografts

Several cadaveric allografts are presently available. Because these grafts have been minimally altered from the initial starting material, they are classified as "minimally processed human tissue" by the FDA. This is an important distinction from the heterografts, which are classified as medical devices and are under closer scrutiny by the FDA. Tissue banks typically regulate these allografts. **AlloDerm®** (LifeCell Corporation, Branchburg, NJ) is created from cadaveric skin using proprietary processing techniques that reportedly maintain the biochemical and structural components of the extracellular matrix promoting tissue regeneration. Cells are then removed by deoxycholate, and the residue is washed and lyophilized. The remnant material is an insoluble matrix composed mainly of collagen, elastin and laminin and closely resembles the composition of normal skin connective tissue (Penttinen & Grönroos, 2008). The graft is noncrosslinked, and is freeze dried, and requires a 20 to 30 minutes soak before use. The ability of AlloDerm to withstand hostile environments has been well documented. However, the graft's durability and the prevention of hernia recurrence have been less clear. **AlloMax®** (Tutogen Medical Inc., Alachua, FL) is another acellular human dermal product that is marketed through Davol Inc. It is noncrosslinked and undergoes a proprietary processing developed by Tutoplast similar to the Tutopatch previously described. **Flex HD®** is manufactured by Musculoskeletal Tissue Foundation (Edison, NJ) and is distributed by Ethicon Inc. (Somerville, NJ). It is stored in a 70% ethanol solution and remains in a hydrated form and therefore does not require rehydration before use. It does not require refrigeration (Rosen, 2010).

Implantation of a **fibroblast growth factor (bFGF)**-releasing polygalactone polymer rod into the fascial wound of rats has been carried out. This approach reduced the development of primary incisional hernias from 60 to 30%, and recurrent incisional hernias from 86 to 23%. This study also reported that type I collagen staining was significantly increased around the bFGF treated fascia, which was thought to contribute to the results (Dubay et al., 2004).

The use of the patient's hernial sac as biomaterial to correct the hernia and reinforce the surgery has also been described, since it also induces fibroplasia (Silva et al., 2004).

6. Complications from the use of mesh screens

The overall risk of complications reported after inguinal hernia operations varies from 15 to 28% in systematic reviews. The most frequent early complications were hematomas and seromas (8–22%), urinary retention and early pain, and late complications were mainly persistent pain and recurrences. Those are the two most important outcome measures. Chronic pain is an issue that primarily affects patient quality of life, and is the most common complication of otherwise successful inguinal hernia surgery (Perenttinen & Grönroos, 2008). A truly successful hernia repair requires effective bridging or augmentation that will prevent recurrence. If reoperation is required in the event of a recurrence, the incidence of chronic pain increases. Other complications described are foreign body reactions, infection, discomfort, dislocation, migration, erosion and shrinkage of the prosthesis (Junge et al., 2006; Zogbi et al., 2010). The risk of infertility has been considered significant in inguinal mesh operations (Penttinen & Grönroos, 2008). Chronic pain, stiff abdomen, and foreign body sensation are least often observed with the use of a lightweight mesh and a laparoscopic approach (Klosterhalfen et al., 2005). Besides these specific complications caused by the prosthesis, complications common to any surgical procedure, such as respiratory or urinary infection, vomiting, constipation, urine retention, venous thrombosis, hemorrhage, anesthetic complications and even death should not be underestimated. Fortunately, lifethreatening complications were rarely reported (Simons et al., 2009).

7. Measures to avoid complications

First it should be recalled that surgical techniques using mesh result in fewer recurrences than techniques which do not use mesh (Simons et al., 2009). Besides, mesh repair appears to reduce the chance of chronic pain rather than increase it (Collaboration, 2002).

During any surgery, a meticulous, anatomical, precise and aseptic surgical technique should be used. The mesh should be positioned adequately when it is fixed, and it should go 2 cm or more beyond the limits of the margins of existing defects, so that a possible retraction or displacement will not compromise the entire coverage of the hernial defect (Amid, 2001). After all recurrences in humans invariably occurred at the mesh margin, where the mesh interfaced with tissue (Bachman & Ramchaw, 2008). Contact with bowel loops should be avoided, because the adhesions resulting from this contact may cause irreversible damage such as necrosis, digestive fistula and elimination of the material (Mathews et al, 2003). Contact with subcutaneous cell tissue should be avoided to reduce the risk of seroma and infection. It should be placed between two myoaponeurotic layers, not only to avoid contact with the viscerae, or with the subcutaneous, as described above, but also so that the mesh will not fold over and will be directly integrated to these tissues, strengthening them (Falci 1997, 2003).

After surgery, the patient should return gradually to his activities, without intense, abrupt efforts. Risk factors described at the beginning of the chapter should be controlled.

8. The ideal mesh for hernia repair: defining characteristics

Classically the first desirable qualities in the biomaterials described were resistance, durability, good tissue tolerance, flexibility, easy manipulation, non-migration, stability, pervious pores, sterilizability and economic feasibility, besides not producing cysts or malignant changes. All this is still accepted (Cumberland, 1952; Scales, 1953). Other needs were found over time, namely, they should not restrict postimplantation function or future access, they should perform well in the presence of infection and block transmission of infectious diseases, resist shrinkage or degradation over time and be easy to manufacture. (Earle & Mark, 2008). Saberski et al added another characteristic called anisotropy, and they found striking differences between elastic properties of perpendicular axes for most commonly used synthetic meshes (Saberski et al., 2011). From the surgeons' and patients' point of view, the optimal mesh should have minimal adhesion formation, excellent tissue ingrowth with minimal shrinkage, no fistula formation and promote minimal pain and seroma formation. Furthermore, it is important that the mesh causes no change in abdominal wall compliance (Eriksen et al., 2007). The mesh should be flexible but also have a good memory, and it should have elasticity in more than one dimension, allowing it to stretch in more than one direction and then return to its original shape. In this way, the mesh should match the abdominal wall dynamics as closely as possible. Flexibility and memory, which make a mesh more adaptable, are also important to optimize the surgical handling of the mesh. The mesh should have an adequate adhesive quality that requires minimal or no additional fixation, even for large defects. An ideal mesh would be a monofilament mesh that would prevent adhesions yet still enable growth of the adjacent tissue for optimal augmentation (Bringman et al., 2010).

In healthy volunteers, documented measured intra-abdominal pressure via intravesicular measurements was up to 252 mmHg in a variety of maneuvers, including lifting, coughing, and jumping. This correlates to forces of up to 27 N/cm (Cobb et al, 2006). A tensile strength of 16 N was probably more than sufficient to augment the abdominal wall; for bridging of large defects, an increased tensile strength of 32 N may be necessary (Bringman et al., 2010). With these numbers in mind, compare a maximum force on the abdominal wall of 27 N/cm with the measured burst force of some of the more common synthetic mesh materials: Marlex® has a tensile strength of 59 N/cm, Atrium® 56 N/cm, and VyproII® 16 N/cm. Marlex® and Prolene® were both over five times stronger than the calculated abdominal wall strength, and Mersilene® was at least twice as strong (Kinge et al, 1998). A similar trend was noted in an animal study conducted by Cobb. Mesh was implanted into swine for 5 months and then tested for burst strength. Native tissue ruptured at 232 N, LW PP mesh burst at 576 N, MW at 590 N, and HW mesh at 1218 N (Cobb et al, 2006). These data have lent scientific support to the theory that synthetic mesh materials, especially traditional HW PP mesh, are overengineered for their purpose. This excess prosthetic can lead to more complications, including decreased mesh flexibility, loss of abdominal wall compliance, inflammation, and scarring of surrounding tissues, potentially leading to pain, a sensation of feeling the mesh in the abdominal wall, and mesh contraction and wadding, which in turn may result in a recurrent hernia (Bachman & Ramchaw, 2008). Actually, all commercially available synthetic prosthetics today have long-term foreign-body reactions. Given the existing products and body of evidence, the overall density should probably be somewhere between 28 g/m² and 90 g/m² to minimize recurrence and adverse effects of the host foreign-body response. Methods to decrease the density of the prosthetic include reduction in fiber diameter (ie, strength) and number of fibers (ie, increase in pore size). Studies have also shown that a PP mesh with a pore size greater than 600 to 800 mm should result in more of a scar "net" rather than a scar "plate". The "net", compared to the "plate", is less prone to contracture and stiffness of the abdominal wall. Not all small-pore prosthetics are stiff. Consider what is seen clinically with microporous PTFE, and the maintenance of pliability even with encapsulation. It may then be that the architecture (woven versus solid) of the prosthetic is a more significant contributor to performance than the polymer itself. The upper limits of pore size for adequate fixation to prevent recurrence have not been appropriately investigated. Very large pore size (4,000 mm) combined with a partially absorbable component doesn't appear to have any clinical benefits in terms of pain, and may not be sufficient to prevent higher recurrence rates when used with a Lichtenstein technique (Earle & Mark, 2008). An ideal portfolio of meshes would have the benefits of both HW and LW meshes, such as the strength of an HW mesh and the flexibility of an LW mesh with none of the adverse events. The HW microporous meshes have a lower risk of tissue-to-mesh adhesion but carry a risk of encapsulation and foreign body reaction, resulting in decreased integration. LW macroporous mesh results in better tissue ingrowth and lower (ou less) foreign body reaction but may lead to a higher risk of adhesions (Eriksen et al., 2007). A larger pore size also provides optimal flexibility for improved physical properties, allowing a better activity profile post-surgery, but relinquishes memory, which is important for handling during the procedure. A monofilament mesh with a pore size of > 2.5 mm seems ideal. In all hernia repair techniques, a strong mesh is important for augmentation of the abdominal wall and to prevent recurrences (Bringman et al., 2010).

9. Charity campaigns involving biomaterials for low income patients.

Publications on campaigns performed in Africa show encouraging results with the use of sterilized mosquito net mesh. Clarke et al. report their results, implanting sterilized polyester mosquito net mesh in 95 poor patients in Ghana, with 2% infection and no recurrences. They concluded that PE mosquito net mesh is a cost-effective alternative to commercial mesh for use in inguinal hernia repair in developing countries (Clarke et al., 2009). Optimistic results were also described in a study performed before this one, in Burkina Faso, using Nylon (100% Polyamide 6-6) mosquito net mesh, this time describing the complete absence of infection (Freudenberg et al., 2006).

10. Establishing animal models for the development of biomaterials

Animal models resembling the human hernia are a useful tool for researchers to investigate hernia treatment options. The current animal models used to study hernia repair are not perfect. Artificially created hernias in animals are poor hernia models as they do not truly recreate the biological defects that cause hernias, such as collagen defects. Furthermore, the defects that are created to test mesh products are not real-life defects that surgeons would encounter. In order to serve as a useful model, the pathology in the animal must be similar to the human hernia equivalent. One factor when considering an animal model is similarities in the elasticity of the abdominal wall. Although there is no consensus for the most appropriate test or animal model, animal models are useful when comparing different meshes in the same species either in vivo or ex vivo. Studies in humans and large animals are the only way that most issues, such as elasticity, chronic pain, foreign body reaction, and adhesion, will be observed. What problems animal models can solve and which animals are the most appropriate for use will differ depending on the purpose of the study. Small animals or even cell cultures are instructive for studying the inflammatory reaction and biocompatibility, but for abdominal wall function and elasticity, larger animals are more suitable. Although ineffective for other comparisons, pigs are useful to simulate mesh implantation within the human body as pigs have a similar body size to humans. Sheep and rabbits are reasonable models to mimic vaginal operating conditions; potentially, they are also useful models for pelvic floor damage due to pregnancy and birth (Penttinen & Grönroos, 2008; Bringman et al., 2010).

11. Conclusion

Operation techniques using mesh result in fewer recurrences than techniques which do not use mesh.

12. Acknowledgment

The author declares no conflict of interest; none of the mesh manufacturers was involved in any way in this chapter. He particularly thanks three Brazilian herniologists who have guided him throughout his career: Prof. Antonio Portella, Prof. Fernando Pitrez and Prof. Manoel Trindade.

13. References

Abaza, R.; Perring, P. & Sferra, J.J. (2005). Novel parastomal hernia repair using a modified polypropylene and PTFE mesh. *J Am Coll Surg.*; 201(2): 316-17.

Acquaviva, D.E. & Bourret, P. (1944). Cure d'une volumineuse éventration par plaque de nylon. *Bull Méd Soc Chir Marseille*; 5: 17.

Agresta, F. & Bedin, N. (2008). Transabdominal laparoscopic inguinal hernia repair: is there a place for biological mesh? *Hernia*, 12: 609–612 DOI 10.1007/s10029-008-0390-0

Annibballi, R. & Fitzgibbons, J.R. (1994). Prosthetic material and adhesions formation. In: Arregui ME, Nagan RF: in Inguinal hernia: advances or controversies? Oxford: Radcliffe; pp. 115-124.

Adler, R.H. & Firme, C.N. (1959). Use of pliable synthetic mesh in the repair of hérnias and tissue defects. *Surg Gyn Obst*; 108: 199-206.

Amid, P.K. (1997). Classification of biomaterials and their related complications in abdominal wall hernia surgery. *Hernia* 1:15–21

Amid, P.K. (2001). Polypropylene prostheses. In R. Bendavid (Ed), Abdominal Wall Hernias. New York: Springer- Verlag Inc, 2001, Pp. 272-277.

Bachman, S. & Ramshaw, B. (2008). Prosthetic Material in Ventral Hernia Repair: How Do I Choose? *Surg Clin N Am,* 88: 101–112

Bellón, J.M. (2009). Implicaciones de los nuevos diseños protésicos de baja densidad en la mejora de la reparación de defectos herniarios. Revisión de conjunto. *Cir Esp.* 85(5) :268–273

Blatnik, J.; Jin, J. & Rosen, M. (2008). Abdominal hernia repair with bridging acellular dermal matrix — an expensive hernia sac. *Am J Surg* 196:47–50

Bringman, S.; Conze, J.; Cuccurullo, D.; Deprest, J.; Junge, k.; Klosterhalfen, B.; Parra-Davila, E. Ramshaw, B. & Schumpelick, V. (2010). Hernia repair: the search for ideal meshes. *Hernia* 14:81–87. DOI 10.1007/s10029-009-0587-x

Brown, J.B.; Fryer, M.P. & Ohlwiler, D.A. (1960). Study and use of sinthetic materials such as silicones and teflon, as subcutaneous prostheses. *Plast Reconst Surg*; 26: 264-279.

Burger, J.W.; Luijdendijk, R.W.; Hop, W.C.; Halm, J.A.; Verdaasdonk, E.G. & Jeekel, J. (2004). Long-term follow-up of a randomized controlled trial of suture versus mesh repair of incisional hernia. *Ann Surg* 240:578–585

Castro, C.C. & Rodrigues, S.M.C. (2007). Cicatrização de feridas. In Programa de atualização em cirurgia. Porto Alegre. Artmed, 2007; 2(4): 35-50.

Clarke, M.G.; Oppong, C.; Simmermacher, R.; Park, K.; Kurzer, M.; Vanotoo, L. & Kingsnorth, A.N. (2009). The use of sterilised polyester mosquito net mesh for inguinal hernia repair in Ghana. *Hernia* 13:155–159 DOI 10.1007/s10029-008-0460-3

Chan, G. & Chan, C.K. (2005). A review of incisional hernia repairs: preoperative weight loss and selective use of the mesh repair. *Hernia*; 9: 37-41.

Cobb, W.S.; Burns, J.M.; Peindl, R.D. (2006). Textile analysis of heavy weight, mid-weight, and light weight polypropylene mesh in a porcine ventral hernia model. *J Surg Res*, 136(1):1–7.

Collaboration, E.H. (2002). Repair of groin hernia with synthetic mesh: meta-analysis of randomized controlled trials. *Ann Surg* 235:322–332

Cumberland, V.H. (1952) A preliminary report on the use of prefabricated nylon weave in the repair of ventral hernia. *Med J Aust* 1:143–144

Deprest, J.; De Ridder, D. & Roovers, J.P. et al (2009). Medium term outcome of laparoscopic sacrocolpopexy with xenografts compared to synthetic grafts. *J Urol* 182:2362–2368

Dubay, D.A.; Wang, X.; Kuhn, M.A.; Robson, M.C. & Franz, M.G. (2004). The prevention of incisional hernia formation using a delayed-release polymer of basic fibroblast growth factor. *Ann Surg* 240:179–186.

Earle, D. & Romanelli, J. (2007). Prosthetic materials for hernia: what's new. *Contemporany Surgery*; 63(2): 63-69.

Earle, D.B. & Mark, L.A. (2008). Prosthetic material in inguinal hernia repair: How do I choose? *Surg Clin N Am*; 88:179–201.

Eriksen, J.R.; Gogenur, I. & Rosenberg, J. (2007). Choice of mesh for laparoscopic ventral hernia repair. *Hernia* 11:481–492. DOI 10.1007/s10029-007-0282-8

Falci, F. (1997) Análise crítica das próteses na região inguinal. In: Silva, A.L. Hérnias da parede abdominal. Clin Bras Cir CBC. Rio de Janeiro, Atheneu, 1997 ano III. V.1, p 141-151.

Falci, F. (2003) Utilização das próteses em hérnias da parede abdominal. In: Lira, O.B. & Franklin, R. *Hérnias – Texto e Atlas*. 1ª ed, Rio de Janeiro, Rubio, 2003, pp. 199-202.

Franklin, R.; Lira, O.B. & Filho, A.M. Introdução ao estudo das hérnias. In: Lira, O.B. & Franklin, R. *Hérnias – Texto e Atlas*. 1ª ed, Rio de Janeiro, Rubio, 2003, pp. 1-18

Gianlupi, A. & Trindade, M.R.M. (2004) Comparação entre o uso de fio inabsorvível (polipropileno) e fio absorvível (poliglactina 910) na fixação da prótese de polipropileno em correção de defeitos músculo-aponeuróticos da parede abdominal. Estudo experimental em ratos. *Acta Cir Bras*; 19(2): 94-102.

Goepel, R. (1900). Über die Verschliessung von Bruchpforten durch Einheilung geflochtener, fertiger Silberdrahtnetze. Gesellschaft für Chirurgie XXIX Congress 1900, 174-177.

Goldstein, H.S. (1999). Selecting the right mesh. *Hernia*. 3:23-26.

Gonzalez, R.; Fugate, K.; McClusky, D.; Ritter, E.M.; Lederman, A. & Dillehay, D. (2005). Relashionship between tissue ingrowth and mesh contraction. *World J Surg*; 29: 1038-1043.

Handley, W.S. (1963). En: Behandlung der Bauchnarbenbrüche. Reitter, H. *Langenbecks Arch Klin Chir*; 304: 296-297.

Hollinsky, C.; Sandberg, S.; Koch, T. & Seidler, S. (2008). Biomechanical properties of lightweight versus heavyweight meshes for laparoscopic inguinal hernia repair and their impact on recurrence rates. Surg Endosc (2008) 22:2679–2685. DOI 10.1007/s00464-008-9936-6

Hutchinson, R.W.; Chagnon, M. & Divilio, L. (2004). Pre-clinical abdominal adhesion studies with surgical mesh. Current Issues Technology. Business Briefing: Global Surgery: 29-32.

Junge, K.; Klinge, U.; Rosch, R.; Klosterhalfen, B. & Schumpelick, V. (2002). Functional and morphologic properties of a modified mesh for inguinal hernia repair. *World J Surg* 26:1472–1480

Junge, K.; Rosch, R. & Klinge, U. (2006). Risk factors related to recurrence in inguinal hernia repair: a retrospective analysis. *Hernia* 10:309–315

Klinge, U.; Klosterhalfen, B.; Conze, J.; Limberg, W.; Obolenski, B. & Ottinger, A.P. (1998). Modified mesh for hernia repair that is adapted to the physiology of the abdominal wall. *Eur J Surg*;164:951–60.

Klinge, U.; Klosterhalfen, B.; Birkenhauer, V.; Junge, K.; Conze, J. & Schumpelick, V. (2002). Impact of polymer pore size on the interface scar formation in a rat model. *J Surg Res*; 103: 208-214.

Klosterhalfen, B.; Junge, K. & Klinge, U. (2005). The lightweight and large porous mesh concept for hernia repair. *Expert Rev Med Devices* 2:103–117

Koontz, A.R. (1948). Premilinary report on the use of tantalum mesh in the repair of ventral hernias. *Ann Surg*; 127: 1079-1085.

Kneise, G. (1953). Erfahrungen und neue Erkenntnise bei der Perlonnetzimplantation. *Centralblatt für Chirurgie*; 12: 506-511.

Lemchen, H. & Irigaray, J.H. (2003). Ferida pós-operatória: conduta na evolução normal e nas complicações. In: Pitrez FAB. Pré e pós-operatório em cirurgia geral e especializada. 2ª ed. Artmed, 2003. Cap. 15. pp 130-136.

Lichtenstein, I.L.; Shulman, A.G.; Amid, P.K. & Montllor, M.M. (1989) The tension-free hernioplasty. *Am J Surg* 157:188–193

Malangoni, M.A. & Rosen, M.J. (2007). Hernias. :In Townsend: Sabiston Textbook of Surgery, 18th ed. Chapter 44. Copyright © 2007 Saunders, An Imprint of Elsevier

Mandl, W. (1962). Diskussionsbemerkung. Osterr Ges für Chirurgie und Unfallheilkunde. Salzburg, 1962, 6-8.

Mathews, B.D.; Pratt, B.L.; Pollinger, H.S.; Backus, C.L.; Kercher, K.W.; Sing, R.F. et al. (2003). Assessment of adhesion formation to intra-abdominal polypropylene mesh and polytetrafluoroethilene mesh. *J Surg Research*; 114: 126-132.

McNealy, R.W. & Glassman, J.A. (1946). Vitallium plates used in repair of large hernia. *Illinois Med Jour* 1946; 90: 170-173.

Melo, R.S.; Goldenberg, A.; Goldenberg, S.; Leal, A.T. & Magno, A. (2003). Effects of polypropylene prosthesis placed by inguinotomy in the preperitonial space, in dogs: evaluation laparoscopic and microscopic. *Acta Cir Bras 2003*; 18(4): 289-296

Monaco, J.L. & Lawrence, W.T. (2003). Acute wound healing an overview. *Clin Plast Surg*. 30 (1): 1-12.

Negro, P.; Basile, F.; Brescia, A.; Buonanno, G.M; Campanelli, G.; Canonico, S.; Cavalli, M.; Corrado, G.; Coscarella, G. & Di Lorenzo, N. et al. (2011). Open tension-free Lichtenstein repair of inguinal hernia: use of fibrin glue versus sutures for mesh fixation. *Hernia* Volume 15, Number 1, 7-14, DOI: 10.1007/s10029-010-0706-8

Nien, Y.D.; Man, Y.P. & Tawil, B. (2003). Fibrinogen inhibits fibroblast-mediated contraction of colagen. *Wound Repair Regen*; 11: 380-385.

Penttinen, R. & Grönroos, J.M. (2008). Mesh repair of common abdominal hernias: a review on experimental and clinical studies. *Hernia*. 12:337–344 DOI 10.1007/s10029-008-0362-4

Ponka, J.L. (1980). Hernia of abdominal wall. WB Saunders Company 1980. Philadelphia.

Rappert, E. (1963). Supramidnetze bei Bauchbrüchen. *Langenbecks Arch Klin Chir*; 304: 296-297.

Read, R.C. (2004). Milestones in the history of hernia surgery: prosthetic repair. *Hernia*; 8(1): 8-14.

Read, R.C. (2011). Crucial steps in the evolution of the preperitoneal approaches to the groin: an historical review. *Hernia*. Volume 15, Number 1, 1-5, DOI: 10.1007/s10029-010-0739-z.

Rodrigues, Jr.A.J.; Rodrigues, C.J.; Cunha, A.C.P. & Yoo, J. (2002). Quantitative analysis of collagen and elastic fibers in the transversalis fascia in direct and indirect hernia. *Rev Hosp Clín Fac Méd S Paulo*; 57(6): 265-270.

Rosen, M.J. (2010). Biologic Mesh for Abdominal Wall Reconstruction: A Critical Appraisal, *The American Surgeon*. Vol. 76, (January 2010), pp 1-6.

Saberski, E.R.; Orenstein, S.B. & Novitsky, Y.W. (2011). Anisotropic evaluation of synthetic surgical meshes. *Hernia*. Volume 15, Number 1, 47-52, DOI: 10.1007/s10029-010-0731-7.

Sans, J.V. (1986). Eventraciones – Procedimientos de reconstrucción de la pared abdominal. 1ª ed. Editorial Jims, 1986. Barcelona España, p 108-116.

Scales, J.T. (1953) Discussion on metals and synthetic material in relation to soft tissues; tissue reactions to synthetic materials. *Proc R Soc Med* 46:647–652

Schönbauer, L. & Fanta, H. (1958) Versorgung grosser Bauchwanddefecte mit implantiertem Chromeatgutnetz. *Brun's Beitr Klin Chir*; 196: 393-401.

Schreinemacher, M.H.F.; Emans, P.J.; Gijbels, M.J.J.; Greve, J.W.M.; Beets, G.L. & Bouvy, N.D. (2009). Degradation of mesh coatings and intraperitoneal adhesion formation in an experimental model. *Br J Surg*; 96: 305–313. DOI: 10.1002/bjs.6446

Silva, H.C.; Silva, A.L. & Oliveira, C.M. (2004). Peritoneum autogenous graft fibroblast: an experimental study. *Rev Col Bras Cir*; 31(2): 83-89.

Simons, M.P.; Aufenacker, T.; Bay-Nielsen, M.; Bouillot, J. L.; Campanelli, G.; Conze, J.; de Lange, D.; Fortelny, R.; Heikkinen, T.; Kingsnorth, A.; Kukleta, J.; Morales-Conde, S.; Nordin, P.; Schumpelick, V.; Smedberg, S.; Smietanski, M.; Weber, G. & Miserez, M. (2009). European Hernia Society guidelines on the treatment of inguinal hernia in adult patients. *Hernia*; 13:343–403. DOI 10.1007/s10029-009-0529-7

Speranzini, M.B. & Deutsch, C.R. (2001). Complicações das Hérnias: Encarceramento/Estrangulamento. In: Tratamento cirúrgico das hérnias das regiões inguinal e crural. 1ª ed. Atheneu, 2001. Cap 4. pp 43-55.

Speranzini, M.B. & Deutsch, C.R. (2001). Biomateriais. In: Tratamento cirúrgico das hérnias das regiões inguinal e crural. 1ª ed. Atheneu, 2001. Cap 5. pp 57-67.

Stoppa, R.G. & Rives, J. (1984). The use of dacron in the repair of hernias of the groin. *Surg Clin North América*; 64: 269-285.

Szczesny, W.; Cerkaska, K.; Tretyn, A. & Dabrowiecki, S. (2006). Etiology of inguinal hérnia: ultrastructure of rectus sheath revisited. *Hernia*; 10: 266-271.

Usher, F.C.; Ochsner, J. & Tuttle, L.L. Jr. (1958). Use of Marlex mesh in the repair of incisional hernias. *Am Surg* 24:969–974.

Usher, F.C. (1962) Hernia repair with Marlex Mesh. An analysis of 541 cases. *Arch Surg*; 84: 73-76.

Witzel, O. (1900). Über den Verschluss von Bauchwunden und Bruchpforten durch versenkte Silberdrahtnetze (Einheilung von Filigranpelotten). *Centralblatt für Chirurgie*; 10: 258-260.

Wolwacz, Jr.I.; Trindade, M.R.M. & Cerski, C.T. (2003). O colágeno em fáscia transversal de pacientes com hérnia inguinal direta submetidos à videolaparoscopia. *Acta Cir Bras*; 18(3): 196-202.

Zogbi, L.; Portella, A.O.V.; Trindade M.R.M. & Trindade, E.N. (2010). Retraction and fibroplasia in a polypropylene prosthesis: experimental study in rats. *Hernia* 14:291–298 DOI 10.1007/s10029-009-0607-x

Permissions

The contributors of this book come from diverse backgrounds, making this book a truly international effort. This book will bring forth new frontiers with its revolutionizing research information and detailed analysis of the nascent developments around the world.

We would like to thank Prof. Rosario Pignatello, for lending his expertise to make the book truly unique. He has played a crucial role in the development of this book. Without his invaluable contribution this book wouldn't have been possible. He has made vital efforts to compile up to date information on the varied aspects of this subject to make this book a valuable addition to the collection of many professionals and students.

This book was conceptualized with the vision of imparting up-to-date information and advanced data in this field. To ensure the same, a matchless editorial board was set up. Every individual on the board went through rigorous rounds of assessment to prove their worth. After which they invested a large part of their time researching and compiling the most relevant data for our readers. Conferences and sessions were held from time to time between the editorial board and the contributing authors to present the data in the most comprehensible form. The editorial team has worked tirelessly to provide valuable and valid information to help people across the globe.

Every chapter published in this book has been scrutinized by our experts. Their significance has been extensively debated. The topics covered herein carry significant findings which will fuel the growth of the discipline. They may even be implemented as practical applications or may be referred to as a beginning point for another development. Chapters in this book were first published by InTech; hereby published with permission under the Creative Commons Attribution License or equivalent.

The editorial board has been involved in producing this book since its inception. They have spent rigorous hours researching and exploring the diverse topics which have resulted in the successful publishing of this book. They have passed on their knowledge of decades through this book. To expedite this challenging task, the publisher supported the team at every step. A small team of assistant editors was also appointed to further simplify the editing procedure and attain best results for the readers.

Our editorial team has been hand-picked from every corner of the world. Their multi-ethnicity adds dynamic inputs to the discussions which result in innovative outcomes. These outcomes are then further discussed with the researchers and contributors who give their valuable feedback and opinion regarding the same. The feedback is then collaborated with the researches and they are edited in a comprehensive manner to aid the understanding of the subject.

Apart from the editorial board, the designing team has also invested a significant amount of their time in understanding the subject and creating the most relevant covers. They scrutinized every image to scout for the most suitable representation of the subject and create an appropriate cover for the book.

The publishing team has been involved in this book since its early stages. They were actively engaged in every process, be it collecting the data, connecting with the contributors or procuring relevant information. The team has been an ardent support to the editorial, designing and production team. Their endless efforts to recruit the best for this project, has resulted in the accomplishment of this book. They are a veteran in the field of academics and their pool of knowledge is as vast as their experience in printing. Their expertise and guidance has proved useful at every step. Their uncompromising quality standards have made this book an exceptional effort. Their encouragement from time to time has been an inspiration for everyone.

The publisher and the editorial board hope that this book will prove to be a valuable piece of knowledge for researchers, students, practitioners and scholars across the globe.

List of Contributors

Marthe Rousseau
Henri Poincaré University, Nancy I, France

Giulio Maccauro, Pierfrancesco Rossi Iommetti, Luca Raffaelli and Paolo Francesco Manicone
Catholic University of the Sacred Hearth Rome, Italy

Ylenia Zambito and Giacomo Di Colo
University of Pisa, Italy

Doina Macocinschi, Daniela Filip and Stelian Vlad
Institute of Macromolecular Chemistry "Petru Poni" Iasi, Romania

Mădălina Georgiana Albu
INCDTP – Leather and Footwear Research Institute, Bucharest, Romania

Irina Titorencu
Institute of Cellular Biology and Pathology "Nicolae Simionescu", Bucharest, Romania

Mihaela Violeta Ghica
Carol Davila University of Medicine and Pharmacy, Faculty of Pharmacy, Bucharest, Romania

Ali Demir Sezer
Faculty of Pharmacy, Marmara University, Turkey

Erdal Cevher
Faculty of Pharmacy, Istanbul University,Turkey

Mark A. Suckow
University of Notre Dame, USA

Rae Ritchie and Amy Overby
Bioscience Vaccines, Inc., United States of America

Ana Colette Maurício, Andrea Gärtner, Tiago Pereira and Ana Lúcia Luís
Centro de Estudos de Ciência Animal (CECA), Instituto de Ciências e Tecnologias Agrárias e Agro-Alimentares (ICETA), Universidade do Porto (UP), Instituto de Ciências Biomédicas Abel Salazar (ICBAS), UP, Portugal

Paulo Armada-da-Silva, Sandra Amado and António Prieto Veloso
Faculdade de Motricidade Humana (FMH), Universidade Técnica de Lisboa (UTL), Portugal.

Artur Varejão
Departamento de Ciências Veterinárias, CIDESD, Universidade de Trás-os-Montes e Alto Douro (UTAD), Portugal

Stefano Geuna
Neuroscience Institute of the Cavalieri Ottolenghi Foundation & Department of Clinical and Biological Sciences, University of Turin, Italy

Luciano Zogbi
Federal University of Rio Grande- FURG, Brazil

9 781632 380630